HARVEST

HARVEST

A Year in the Life of an Organic Farm

Nicola Smith with photographs by **Geoff Hansen**

THE LYONS PRESS
Guilford, Connecticut

An imprint of The Globe Pequot Press

Captions:
Page ii: Using a rented tractor and equipment, Kyle Jones cuts grass for hay on 30 acres at Fat Rooster Farm in Royalton, Vermont. To the left is one of the birdhouses for bluebirds and tree swallows on the property.

Page vi: On a warm afternoon, a lamb takes a peek at a visitor while eating hay at the farm.

Page x: Kyle pulling his three-year-old son, Brad, on a sled in the snow.

The Lyons Press is an imprint of The Globe Pequot Press

10 9 8 7 6 5 4 3 2 1

Printed in China

Designed by LeAnna Weller Smith

ISBN 1-59228-234-2

Library of Congress Cataloging-in-Publication Data is available on file.

For Jennifer, Kyle and Brad.

CONTENTS

Acknowledgments

A debt of gratitude is owed to the many people who assisted us in the research and writing, and photographing, of this book. First and foremost, our deep appreciation to Jennifer Megyesi and Kyle Jones for their interest in seeing the project through, their patience, their willingness to answer any and all questions, day or night, their unfailing generosity, their good humor, and last but not least, their integrity. Our thanks also to Louis and Beverly Megyesi and Sandy and Lois Jones, who were never less than helpful. The author would particularly like to thank Betty Morse, who was most informative in giving a history of the farm from 1872 on; she was always a pleasure to talk to, and scrupulous in her attention to historical detail.

Numerous individuals helped in ways large and small, typically under tight deadlines and with the utmost professionalism and courtesy. They are: Enid Wonnacott of NOFA-Vermont; Pat Joyce and Daryl Brinkman of the National Agricultural Statistics Service, United States Department of Agriculture; Steven F. Justis and Curtis Stasheski of the Vermont Agency of Agriculture; Holly Givens and Katherine Dimatteo of the Organic Trade Association; Dan Billin at the Valley News in Lebanon, New Hampshire; Franny Eanet of the Norwich Farmers' Market; and Jeff Nichols and Nick Greeno of Fresh Farms in Rutland, Vermont.

We would also like to express our appreciation to our editor, Ann Treistman, for her unflagging enthusiasm and support for the project, and meticulous attention to both text and photography, and our thanks to Jane Crosen, our punctilious copy editor, who was alert and sensitive to all matters factual, grammatical and stylistic. All books accumulate a host of such debts, and this one is no different.

Finally, to all our friends and family, and to our three-year-old daughter, Emma, who at the beginning of the project told us that she didn't want to go into the barn because the animals were too noisy, especially the pigs, but now jumps at the opportunity to visit the people and animals at Fat Rooster Farm.

In the spring of 2002, the local paper where I live ran a story about a butcher named Mark Durkee, who goes from farm to farm to slaughter livestock. There are not many such itinerant butchers left in this part of central Vermont, halfway between New Hampshire's western border and Montpelier, Vermont's capital; most farmers ship their livestock to slaughterhouses if they have them in quantity.

At the Norwich Farmers' Market in Norwich, Vermont, Jennifer Megyesi answers a customer's question. Starting the first weekend in May, the market is so popular that there's a two-year waiting list for farmers to get a booth.

But these were not herds of animals that Durkee was slaughtering. A farm family—including Fat Rooster Farm, the subject of this book—might call on him to kill a steer or a pig or lambs that they had raised for meat. He'd shoot and dress them, turn them, essentially, from what had been an animal into meat, and the family would put the meat in a freezer to take them through the winter. His prices were good and he was a personable, chatty fellow who liked to tell stories while he worked, and would buttonhole you on the question of your religious beliefs if he had the chance, and if you let him.

He'd been butchering since he was a young man, and was now approaching seventy. He was considered by many to be one of the most experienced and best butchers they knew. He had the gift of approaching animals without causing them undue fear or stress. He took his time, he was unobtrusive, he didn't fluster easily, and he killed decisively and accurately. One pop of his 0.22 magnum rifle—he usually needed no more than one shot because his aim was unerring, and, in fact, he loaded only one cartridge at a time into the rifle—and the animal crumpled and gave up life.

With the story, the paper ran accompanying photographs: one that showed Durkee dressing a dead steer, peeling away its hide to reveal the flesh and muscle beneath, and a second picture that recorded him shooting a different steer. He took aim, crouched slightly so that he seemed to be at eye level with the animal, looking as intent and serious as a sharpshooter in a Civil War photograph by Matthew Brady or Alexander Gardner. The picture was taken at the decisive moment when the animal was vaulted into the air by the impact of the bullet. The steer looked as though it was levitating, with all four feet off the ground, legs pointed straight down.

Because this is a rural area, with families still farming, although they are fewer and farther between, you might think that the story and pictures would not have excited much controversy. They recorded an inescapable fact about farming, that animals are killed for meat. But, in fact, the story caused a small uproar. There was a flurry of letters to the editor, accusing the newspaper of poor editorial judgment and tastelessness and sensationalism. There were letters that complained of cruelty to animals and letters that blamed the butcher for doing work that was, they said, little short of barbaric.

While Bob the Suffolk Punch draft horse watches with interest, Jennifer spreads grain on the ground for the other animals.

The reaction seemed to be a classic instance of the outsider or flatlander, to use the regional vernacular, bringing with him or her judgments that showed little understanding of what it means to live in a rural place, and even less understanding of rural, or farm, work. No farmer would have written such a letter because a farmer would know that if you are going to kill an animal, it is preferable to do it in the place that it knows best. The animal doesn't know what is about to happen, and it is not agitated or made afraid or injured by the process of removing it from its home and transporting it dozens or hundreds of miles to slaughter.

Intellectually the writers of the angry letters knew that animals are raised and killed for meat, but something that was largely invisible, something that happened on a farm or in a slaughterhouse, had been made visible, and it was unpleasant and perhaps even shocking to them. Although a farmer may form some kind of attachment to his or her animals, they are not pets but livestock whose sole purpose is to feed and clothe us, and to breed more livestock.

The intensity of the public reaction—which, initially, slanted almost exclusively toward empathy for the animal, and none for the butcher, whose livelihood this was, or the animal's owner, who planned to feed his family with it—seemed emblematic of the profound disconnect between human beings and the food we eat and, by extension, the land on which that food is produced.

It would not have surprised me to learn that many of the same people who dashed off indignant letters about the cruel fate of the farm animals were meat eaters who would think nothing of picking up a few pounds of ground chuck, wrapped in cellophane, at the local supermarket. They would not have given thought to where that meat originated, how it had been produced and processed, and how it had come to be there. In most cases, the packaging wouldn't give them any clue as to where the meat had come from or where it had been processed. It could have come from the West, the Midwest, the South, or the Northeast. It could be days old, or weeks old. It could be the meat from a steer, or just as likely, a dairy cow that had outlived its usefulness as a milk producer and been "culled"—selected for slaughter—and shipped to market. At this stage of the game, the meat bore no resemblance, in the mind's eye, to the animal from which it had come.

We know, or think we know, where our food comes from. Food comes from farms. But what does this really mean to people? What should it mean? And why should we care?

Is it surprising that there is such widespread ignorance of how food is produced when less than 1 percent of the American population—or roughly 2.1 million out of 292 million people—is engaged in agricultural work? This is a remarkable decline from 1900, when 38 percent of the American population derived its income from farming. (The census figures show a steady decrease: 1900, 38 percent; 1920, 27 percent; 1940, 18 percent; 1950, 12.2 percent; 1960, 8.3 percent; 1970, 4.6 percent; 1980, 3.4 percent; 1990, 2.6 percent; 2002, 0.8 percent. This decrease is even more striking when measured against the increase in population, from nearly 76 million in 1900 to 292 million in 2004.)

The discipline of agriculture—a body of knowledge that used to be known to many—is now as mysterious to us as the calculus required to send rockets into orbit. And as remote: People in cities like New York, Boston, San Francisco, Los Angeles, Dallas, and Chicago are enormously knowledgeable about their food when it arrives at the table, able to speak authoritatively on the virtues of artisanal cheeses from Vermont or Wisconsin or the South of France, heirloom fruits and vegetables grown in Napa and Sonoma, truffles or prosciutto flown in from Italy, and quality meats raised in the Hudson Valley—and yet almost totally ignorant of what is required to grow that food, and what terms like "organic," "natural," or "sustainable" mean in practice. The act of eating food is now almost completely divorced from the act of growing it.

Of course, the demographics of the United States have altered radically in the 104 years since the turn of the twentieth century. Power and wealth are no longer concentrated in land, or the holding of land, but in cities or suburbs, and in the commerce and manufacturing found there. The history of the American Century, particularly in the Depression years and the war and postwar years, has been one of tides of migration from the rural areas to the cities in search of work. This shift is precisely the opposite of what Thomas Jefferson had in mind when he argued that "The United States . . . will be more virtuous, more free and more happy employed in agriculture than as carriers or manufacturers."

Jennifer weighs green beans.

For all his eloquence, Jefferson was a bit monomaniacal on the subject, as when he wrote in a 1787 letter to the Virginia jurist, John Blair, that the "pursuits of agriculture [are] the surest road to affluence and best preservative of morals." One might make a case for the latter, although farmers are no better or worse as humans than anybody else, but as for the pursuit of agriculture leading to affluence, that was an optimistic prognostication which most small farmers—perpetually struggling with financial uncertainty—would have a hard time crediting, even, I think, in Jefferson's time.

Yet the image of the small, family American farm is as iconic today as it was when Jefferson insisted that America's commercial power—indeed, its moral superiority—as a nation would arise from its agricultural base, and that prosperity and civility and civic virtue were the natural, inevitable outcome of yeoman farmers cultivating their gardens.

That agrarian ideal still exerts a powerful grip on the American imagination, even though our actual experience with it may extend no further than mouthing the words to "Old McDonald Had a Farm." (I've been struck, reading to my three-year-old daughter, how odd it is that most American children—the majority of whom are raised in cities or suburbs, not on farms—are still reared on books in which animals baah, moo, cluck, or neigh.)

Like Jefferson, we impute virtue to farming, and we mythologize it in a particularly American way, as the little guy, the underdog, holding on in the face of corporate greed and governmental indifference because of his or her attachment to The Land. We romanticize it, although much of American agriculture is a business like any other, propped up by subsidies and tariffs and lobbying, and the history of the independent American farmer in the twentieth century has been, to say the least, turbulent.

We imbue the farmer's sufferings—the failure rates, the bankruptcies, the accidents—with a tinge of stubborn nobility, as if the worthiness of the endeavor made its perceived demise that much more tragic. Every year since 1985, singers like Willie Nelson, John Mellencamp, and Neil Young have put on a benefit called Farm Aid, whose proceeds are intended for farmers in dire financial need. It would be hard to imagine the same set of celebrities putting on a benefit called, say, Steel Industry Aid, or Rock Stars for the Rust Belt. Decrepit factories and unemployed mill workers, although poignant in their own where-did-the-American-dream-go way, don't have the same aura of small-town whole-someness or purity as the small farmer, to whom the adjective "beleaguered" is almost invariably attached, a description that while perhaps financially accurate smacks of unwitting condescension.

We tend to think of small farmers in a dreamy, nostalgic haze which is, in a sense, to declare them extinct, part of our agrarian past but not our future. Their existence becomes more in the nature of a quaint curiosity from an earlier era, like the trolley car or the phonograph, and not something vital and living. Yes, the number of small farms has declined for a variety of complex reasons, but there are still farmers out there, many of whom might say that accounts of their death have been exaggerated.

The purpose of this book is to look at some of the issues that surround farming, and the question of where our food comes from, by examining one farm, Fat Rooster Farm, in one town, Royalton, in one state, Vermont, during one year's time, 2003.

Fat Rooster Farm has been owned and run by Jennifer Megyesi, forty-one, and her husband, Kyle Jones, forty-four, since 1998. They have a son, Bradford Kipling Jones, who is four. Fat Rooster Farm is an organic farm, which means that Megyesi and Jones are not reliant on any kind of chemical intervention, either with the animals or with the produce they grow or with the land. No fertilizers, no genetically modified seeds, no pesticides.

In one sense, Jennifer Megyesi and Kyle Jones represent the new face of farming: they're younger, in their early forties; they're college educated and have both earned master's degrees; they've worked off the farm, and both continue to rely substantially on income derived from working at jobs off the farm; they're better traveled; and they've been influenced by the environmental and social movements of the 1960s and 1970s. Their life together revolves mostly, but not entirely, around farm life, which makes them quite different from farmers of the previous generation who counted themselves lucky if they took a vacation once every seven years.

In another sense, the new face of farming as embodied in Fat Rooster Farm is the old face revised and updated, the kind of farming we think about when we think about farming: sheep out at pasture,

Kyle picks sweet corn. He doesn't allow anyone else to pick the corn, because he says he knows when it's ready by picking an ear and eating it for breakfast.

chickens that roam free, cows trailed by their calves, hand-gathered honey and maple syrup, and a farmer kneeling in the earth and bringing back up handfuls of the richest, darkest loam.

Fat Rooster Farm is deliberately small, both in size (twenty acres) and in intention, and it is run by Megyesi and Jones alone, albeit with the help of farm apprentices who sign on for the summer or for a short spell in the winter or spring. They don't want to get too big and bring on the kinds of problems—namely, debt—that they think have plagued other farmers. It is an economy of scale.

They grow enough to sell and, as importantly to them, they grow enough to feed themselves—their own meats, eggs, fruit, honey, and vegetables. They have rejected the central tenets of what is called conventional farming: they don't take out loans for equipment; they barter services and equipment with neighbors; they don't use pesticides or herbicides or chemical fertilizers; and they don't engage in monoculture, which is the same crop planted year after year in the same spot.

In particular, what distinguishes Jones and Megyesi from "conventional" farmers in Vermont, which is still primarily a dairy state, is that they don't do dairy—which may, in fact, be their salvation, given the precipitous speed at which small Vermont dairy farms, the kind that milk twenty to sixty cows, go under yearly. Unlike Vermont's conventional dairy farmers, who are at the mercy of prices set elsewhere for the milk they produce and the grain they buy and the cost of transporting that milk to market, Jones and Megyesi have attempted, inasmuch as it is possible, to eliminate the middleman. They endeavor to find their own markets and set their own prices and grow their own forage for their animals. They also let the market come to them, in the form of Community Supported Agriculture (CSA), in which people who want organic meat and produce pay Fat Rooster Farm a set amount for a six-month season and in return get a certain amount of food each week. They cobble together a modest living that, while not substantial, affords them enough income to contemplate doing it all over again, with adjustments here and there, the next year.

The story of Fat Rooster Farm is one of resources, and a marriage, occasionally stretched to the breaking point but not, in the end, broken definitively. There is life and death in equal measure, of an intensity and frequency with which very few of us are acquainted. Animals are born, and it is astonishing to see it. And then those very same animals that you shepherded into the world, you shepherd out again, when you kill them—and that is imbued with its own weight.

Fat Rooster Farm is, admittedly, a microscopic part of the enormous whole that is American agriculture. In 2004, only approximately 7,000 of the 2.1 million farmers in the United States are certified organic, according to the United States Department of Agriculture. This is minuscule, but in fact, the organic market in this country is estimated to have an annual growth rate of as much as 20 percent, making it one of the fastest-growing segments of American agriculture. However, that number may vary depending on which sector of organic farming we are talking about (produce or livestock or fabrics); also one has to keep in mind that a 20 percent growth rate represents a sector that is still only 1 to 2 percent of the whole.

The number of acres in organic production, both pasture and cropland, jumped by nearly a million acres in forty-eight states from 1997 to 2001: from 1,346,558 in 1997 to 2,343,857 in 2001, or roughly 0.3 percent of farmland. In 2002, North America went ahead of Europe as the largest market for organic goods, spurring an approximate 10 percent increase in the organic market globally. The organic market accounts for 1 to 2 percent of total food sales in the U.S. (not all organic products are American made, however), and in another decade it's estimated that, if it continues at its present rate, perhaps some 2.75 percent of farmland will be certified organic acreage. It's estimated that organic farmers may receive as much as 20 percent more for their goods than conventional farmers, although that figure represents gross, not net, sales, and therefore does not take into account the higher cost and labor of producing organic foods.

In short, while organic goods—and the farmers who produce them—represent no more than a narrow slice of a large pie, the slice will, if projections hold, get bigger. Organic farming is not a passing whim, no longer simply a hangover from the back-to-the-land movement of the 1970s, but an industry that has found a significant toehold.

Consumers are attracted to organic produce for any or all of the following reasons. They think organically grown food is safer for them and their children. It has the stamp of better

Organic eggs, syrup, and herbs are displayed at the Fat Rooster Farm booth, for sale at the farmers' market.

quality—fresher, locally grown, "artisanal" food. And it seems to offer environmental benefits—preservation of farmland and sound, ecologically driven principles of land management. In a country in which cheap food, like cheap gas, is considered something of a national birthright, organic food has the cachet of superior quality, and people are willing to pay for it for reasons that are as much intellectual and philosophical (perceived environmental benefits) as they are concrete (fresher food).

Indeed, it is a measure of the consumer demand for organic foods that large corporations like Purdue and Tyson, sniffing profit, have entered into the organic market—a donning of the organic mantle that the small farmer tends to regard with a not unnatural cynicism. It is also an indication of the significant growth of organic farming that in October 2002, the United States Department of Agriculture (USDA)—twelve years after the federal government set standards for the production, processing, and certification of organic food in the Organic Food Production Act of 1990—finally implemented a comprehensive national policy under the direction of the National Organic Standards Board, a wing of the National Organic Program. Any food labeled organic must now meet the national organic standard.

Until this implementation, nongovernmental organizations had acted as certifiers, or watchdogs, and any compliance with an organic standard by a farm or business was not only voluntary, but piecemeal—a tangle of local and state regulations that were bewildering to the consumer, if the consumer even bothered to look into them (a big "if"). In other words, you could buy a head of broccoli that was labeled "organic" and, unless you personally knew the producer, or were familiar with the farm's methods, you would have to take it on trust that that was the case.

Now an organic producer can affix a USDA seal of organic certification to the product—use of the seal is voluntary—and the consumer can, presumably, be confident that the seal stands for something. Of course, in the minds of many small farmers, for whom any governmental regulation is suspect, the USDA is perceived as a not entirely benign or evenhanded overseer, but as a bureaucracy that can always be counted upon to end up in the pockets of agribusiness.

Organic farming may never approach corporate farming in terms of scale or significance or political influence, and it cannot be called a panacea for some of the problems that trouble American agriculture as a whole. But it seems to be a way to preserve pockets of rural land that might otherwise be folded under for development; and, because it boosts farm income by commanding higher prices than conventionally grown products, it offers people who want to farm a way to make a decent enough living if they're canny about it.

Farming of this kind has one foot in the harder-edged realities of the business world (even if in idealistic opposition to many of those realities), and the other in something much harder to define and describe without lapsing into the kind of phony, amber-waves-of-grain lyricism that could only be written by someone who doesn't have to do it for a living. There is an attachment to the land, an attachment to the animals on the land, the willingness to do the kind of physical labor that most of us couldn't or wouldn't do, the satisfaction of providing the food that people eat, and the gratification of self-sustenance, despite all the attendant financial anxiety and familial strain farming can produce.

What drives and inspires farmers in 2004 to farm—including Jennifer Megyesi and Kyle Jones at Fat Rooster Farm—is not, I think, all that different from the effusive sentiments expressed by J. Hector St. John de Crèvecoeur in his classic 1782 text, *Letters from An American Farmer,* published one year after the Treaty of Paris officially ended the American Revolution. As a new country arose into being, one in which the promise of equal distribution of land was heralded as one of the hallmarks that would forever divide and distinguish the New World from the Old, Crèvecoeur wrote:

"What should we American farmers be without the distinct possession of that soil? It feeds, it clothes us, from it we draw even a great exuberancy, our best meat, our richest drink, the very honey of our bees comes from this privileged spot. No wonder we should thus cherish its possession. . . . This formerly rude soil has been converted . . . into a pleasant farm, and in return, it has established all our rights; on it is founded our rank, our freedom, our power as citizens, our importance as inhabitants of such a district."

Kyle uses a
neighbor's tractor
to smooth the land
with a harrow.

HARVEST

THE FARM

Something was killing the chickens. It crept into the barn in the hours between nightfall and dawn and took the laying hens a few at a time. You could see where it had dragged the dead or dying birds, because of the trail of feathers that lay pell-mell, as though someone had ripped open a down pillow and shaken it vigorously, scattering feathers into the air and leaving them where they fell. There were no tracks, not yet, but the predator's modus operandi seemed fairly clear. It slipped in from the outside

A Golden Comet hen sets on her eggs. The farm's chickens are spread around the property in the summer.

by going under the gutter cleaner, a contraption designed to remove manure from the barn on a conveyor chain, and took an immediate left turn into an area that was set aside for the laying hens. Here the hens—a cacophony of Araucanas, Black Minorcas, White Leghorns, Barred Rocks, and Golden Comets—ate, drank, socialized, squabbled, brooded, and laid their eggs. It was hard to tell whether it was killing the birds there and moving them, or seizing birds that happened to be roosting elsewhere—near the sheep or pig pens, or in what Jennifer called the Old Barn, a cluttered, cobwebby area of boxes and wood and wire mesh and wheelbarrows that was, even in the hottest weather, damp and cool.

It was the height of summer, late July 2003. The predator was determined and crafty and apparently hungry, with the advantages of stealth and darkness on its side. It had come nearly every night for close to twenty-one nights. It had begun by killing a few hens that were incautious enough to move outside at dawn, before daylight swept any nocturnal predators back to their haunts. At first Jennifer had suspected an owl, because of the way the heads were ripped off and because the chickens were free-range and could move around at will, inside or out, leaving them susceptible to a silent, diving attack from the air. (Chickens, although largely flightless, can run with surprising, herky-jerky, silent-film speed, but not quickly enough to escape an owl or a hawk swooping out of the sky.)

But when the killing continued unabated, and the manner of it changed beyond decapitation—the chickens' backs broken and eaten out, the profusion of feathers—Jennifer and her husband, Kyle, began to suspect raccoons. They'd talked to a man who knew and hunted raccoons and he told them, No, raccoons aren't killers, they're scavengers; something else is killing your birds. But Jennifer and Kyle weren't persuaded. Every morning, at daybreak, Jennifer would come down to the barn to do chores, and every morning, there were more dead birds, or what was left of them, scattered on the ground. In five years of farming, Jennifer and Kyle had never had such a serious predator problem and if they couldn't get it, it might well kill every single bird they had.

Whatever it was, was an opportunist. Not content with the chickens, it had gone on to kill poults (young turkeys) kept in the Old Barn, near the stalls that held the two horses, Michael, an

Anglo-Arabian, and Bobby, a Suffolk Punch draft horse. The poults were housed in a large wire-mesh crate, with a waterer and a grain trough. There had been fifty of them to start with, Broad-breasted White Turkeys that had been shipped through the U.S. mail and were picked up at the post office like any other package. They'd arrived in June, one day old, and would be ready for slaughter, at twenty weeks of age, just before Thanksgiving.

Jennifer and Kyle had already suffered previous losses earlier in the summer when, in one of those freakish accidents that can occur in farming, seventeen poults had piled on top of each other to the point of suffocation. The poults were small and white, with pink beaks and bright unblinking eyes. They were spindly and vulnerable looking and curious, with a way of tilting their heads and elongating their skinny necks when they looked at you, as if conveniently offering themselves up for slaughter. They sent up little peeps of alarm when you approached, and for protection massed together in a great feathery ball.

But they were guileless enough that when whatever killed them walked up to their crate, they stuck their heads out through the mesh to see what was going on. At which point the predator decapitated them. That was one theory, anyway, judging by the headless bodies scattered inside the crate. The second theory was that the predator simply reached in between the mesh and swiped at them with a paw. They were, after all, rather like fish in a barrel, and the predator got ten of them. Ten dead turkeys: approximately $600 lost in revenue at Thanksgiving, on top of the losses incurred when the other poults suffocated.

In desperation, Jennifer and Kyle borrowed from a trapper two Conibear traps—the kind that would kill the animal instantly—and two leg-hold traps and placed them carefully at the animal's most likely points of entry, with stern admonitions to their three-year-old son, Brad, to NEVER go near them. But the animal was too clever, and either ignored the traps or set them off, without itself being caught. "The minute we put those things up, it knew," Jennifer said.

After that failure, Jennifer and Kyle instituted a kind of prison lockdown—every entrance and exit, every nook and cranny, that could be sealed, was sealed. Nothing could get in or out. For a while that seemed to work, but when Jennifer left Vermont for a few days at the end of July, to visit her sister, Anna, in West Virginia, she took down the traps because she didn't want to endanger her cats, who had the tendency to wander in and out, and it returned. From early July through the end of the first week in August, the predator had killed thirty-five out of fifty laying hens, and the ten turkeys. This was a significant loss for a small operation like Fat Rooster Farm, and not one that could be easily recouped. And they still didn't know, for certain, what IT was.

Then one day it left tracks. At the entrance to the henhouse, almost directly under a U.S. Fish and Wildlife Service placard that read BIRD NESTING AREA: PLEASE DO NOT PROCEED BEYOND THIS SIGN!, you could, if you bent down and looked closely, see blurred animal prints left behind in the sawdust that was put down to keep the barn smelling relatively sweet. More tellingly, there was the distinctive, unmistakable impression of a raccoon's paw—vaguely webbed and about the size and shape of a baby's foot—in the dust on top of one of the four lidded garbage cans that contained grain and stood directly outside the partition to the henhouse. The raccoon had clambered up onto it to get at the collected eggs that Kyle, whose job it was to gather them, typically left in a hanging basket above the cans. It had broken the eggs, probably with its teeth, sucked out the yolks and whites, and left a debris field of shells behind.

Short of putting in a surveillance camera, this was as telling a confirmation of its identity as you could hope for. But identification was not the same as apprehension. Not only was the raccoon decimating the flock of laying hens and the turkeys, but there was another point of vulnerability for which Kyle had special concern: row upon row of sweet corn down in the fields below the barn, growing fast and thick and tall in the summer heat, and given an additional powerful spur by July's heavy rain. There were people who swore up and down that, when the conditions were right, you could literally hear the corn grow overnight—a rustling sound as if the wind were blowing through the papery stalks—and now the conditions were right.

For raccoons, who love the plump, sugary kernels of sweet corn, stumbling on such a harvest at its peak would be impossible to resist, akin to the career criminal discovering that the vaults at Fort

Jennifer opens the chicken coop door during morning chores. The coop used to be the heifer barn when the farm was used for dairy.

Sweet hand-picked corn is ready to be loaded into the truck for delivery to the South Royalton market. Everyone worried that it was a matter of time before the raccoon that was attacking and killing the farm's laying hens would find and decimate the sweet corn.

Knox have been left unattended. They could easily make short work of the corn crop in one or two nights, which, given the high demand for Fat Rooster corn at farmers' markets and the local co-ops, would be a very bad thing—as if the loss of the hens and turkeys wasn't bad enough.

These are the contingencies for which a farmer may intellectually plan—mortality rates in livestock, a certain percentage of crop loss, vagaries of weather—but when something actually goes wrong, or threatens to go wrong in a hurry, taking a portion of your livelihood with it, it feels as though a solution can't come quickly enough.

So Jennifer took out Kyle's 20-gauge shotgun that was kept in the cellar and went down into the barn and sat in a chair and waited for the raccoon to show itself. It didn't come the first night, and it didn't come the second night when Kyle sat watch, although at one point Kyle thought he saw something furry that might be a raccoon peering into the barn and then retreating when it spotted him. On the third night, the night of August 6, which was clear and balmy, Kyle went down into the barn again. He sat facing the henhouse, arms crossed across his chest, shotgun laid across his knees, legs planted firmly apart, looking like a homesteader in a Western, the kind of man who sits out on his porch waiting for the cattle rustlers to appear on the horizon.

Jennifer and Kyle had already moved the hens out of the henhouse as a precaution, but that had not stopped the killing. Some of the hens had taken to huddling near the cow stalls, under the one bare bulb that was always on at night, taking shelter within that small circle of light, looking around anxiously at any unexpected noise, clucking to themselves. They were being gunned for, and they knew it. Still others roosted up in the haymow—the upper, vaulted part of the barn where baled hay was stored—sending up squawks of warning when a stranger passed by. The barn was empty except for the hens, the pigs, and Blackberry Stem, a Jersey calf the color of café au lait, which was being housed there indefinitely by a neighbor. The sheep were out at pasture, as were the cows.

The stage was set, but the barn couldn't be called quiet. It creaked, the hens muttered, the calf shifted restlessly in its stall, twisting its head around to look at Kyle, and the sows—Cookie, Old Spot, and the truculent sow known as White Pig—vied for the best sleeping spots. Three six-hundred-pound sows in competition for optimum position in one pen, even a large pen, do not go gently into the night. There were sporadic eruptions of testy, ill-tempered, stubborn squealing while the sows changed position.

Kyle stared fixedly at the gutter cleaner, where they had determined the raccoon had been entering and exiting. It was hard to see, straining to look at what could be something or might be nothing, ears sifting through the range of noises that the barn could produce—bangs, clatters, snuffles, sighs, and whispers. Harder still to stay awake when you've been up at dawn, squatting on the ground, harvesting beans and tomatoes and corn and chard and cucumbers and peppers and then taking them to farmers' market.

If he did see the raccoon and he was able to get off a shot in time, it would sound, in this small enclosure, like the roar of a cannon, and scare the hell out of the animals, not to mention humans, which is why he wore the kind of protective earpieces a construction worker uses to muffle noise. The first night, when Jennifer had been leaving her watch, walking the short distance from the barn to the house, she had accidentally discharged the shotgun and that had been loud enough and shocking enough to wake Kyle out of a sound sleep and send him stumbling down the stairs and out into the yard to see whether Jennifer had shot the raccoon or herself. Neither, as it turned out.

On the third night, he sat in the barn for an hour and a half, eyelids drooping lower and lower into sleep, and finally gave up. If it hadn't shown itself just after dusk, it probably wouldn't come again until dawn, its favored hours of predation. Dramatic logic dictated that on the third night the raccoon would appear, and that the shotgun introduced in the first act would be fired by the third, but raccoons don't operate on dramatic, or human, logic, and it clearly knew enough to outwait the human waiting for it. No, if they were going to catch the raccoon, it wouldn't be this night.

But, the next morning, when Jennifer went down to do the chores, she found it: in the henhouse, in a live-catch trap that they'd borrowed from a neighbor and used as a last resort. Jennifer had driven over to Bethel U-Save Discount Grocery and picked up some dented cans of mackerel to use as bait. Wriggling her way in and out of the 4-foot cage, which was not easy to do, even with her small,

Jennifer bottle-feeds Blackberry Stem, a Jersey calf that belongs to a friend who didn't have a barn to house the animal.

The Fat Rooster Farm hens take a drink of water as the morning heat begins to build.

wiry frame, she'd smeared the trap liberally with the mackerel, which smelled to high heaven, and then, to disguise the fact that it was a trap, she'd piled hay high against its sides, leaving nothing exposed but the very top.

And it had worked. Here IT was. Caught.

"Hello, you sonofabitch," Jennifer bellowed. The raccoon had killed thirty-five chickens and ten turkeys and had terrorized the remaining birds, and yet it looked quite small and tense in its cage, gathered tightly into itself, eyes wary, not particularly murderous looking, and unexpectedly and against all reason, she felt sorry for it. But sentimentality is for people who don't farm, and who aren't confronted, as farmers are daily, with the inescapable, inconvenient, messy realities of life and death. There seems to be a general rule of thumb that the farther removed you are from nature and from the necessity of having to earn a living from it, the more sentimental and romantic you are likely to be about it—as if farms should operate on the same principles as a children's petting zoo or a show farm, where the animals are always docile and always smell sweet and never grow to maturity, always stay playful and fleecy and downy and snuggly, and are never slaughtered for meat or because they are injured or sick or worn out or just too damn mean.

It was all well and good to say that the raccoon was just being a raccoon, and following its instincts, and this was part of the larger Darwinian imperative, and let it alone, but it had taken forty-five largely helpless birds with it—birds that, over hundreds of years, had had the ability to fly, to escape, bred out of them by man. The chickens no more wanted to be victims of the raccoon than Jennifer and Kyle wanted them to be. Intervention was required. The raccoon had chased and caught, and now it, in turn, had been chased and caught. If you let it go, even miles away, it would come back to the food source, because clearly it knew a good thing when it saw it, and the killing would start all over again, and this Jennifer and Kyle could not allow. They had invested too much time and money and labor in the chickens to see them picked off one by one.

So they'd made an arrangement: a man who had a permit to hunt raccoon was going to take it miles away and let his dogs get its scent; then he would let the raccoon go in the woods and the dogs would attempt to tree it or run it to earth, and that would likely be the end of the raccoon. There was some kind of rough justice here. In return, the farm was not troubled any further that summer by the raccoon, or any other predator.

Fat Rooster Farm lies at the outer edges of Royalton Township, in Windsor County, in what is called the Upper Valley. The area, which takes its name from the upper valley of the Connecticut River watershed, comprises four counties and forty-seven towns on either side of the river in western New Hampshire and eastern Vermont, and is roughly halfway between the Massachusetts and Canadian borders.

All told, the farm—house and barn, woods, cropland, and pasture—comprises 20 acres. Both the farmhouse and the barn sit on a rise overlooking fields and woods. Below them runs the Second Branch of the White River. (The White River meanders some forty-six miles from its source in Granville, on the eastern slope of the Green Mountains, down to White River Junction, where it joins the Connecticut River, passing the towns of Bethel, Royalton, Sharon, and Hartford on the way. It has three tributaries: the First, Second, and Third branches.) The mill town of Bethel lies some 4 miles to the west of the farm; the town of Randolph, some 12 miles north. Vermont's capital, Montpelier, is a 45-minute drive north; Hanover, New Hampshire, home to Dartmouth College, is 35 minutes south.

Jennifer and Kyle called the farm Fat Rooster after two overweight, white roosters, Forrest Gump and Wendell Berry, that were Cornish-Rock crosses. Kyle disliked these birds, or rather, he disapproved of the way Jennifer kept trying to turn some of the birds and animals into pets. The minute you turned livestock into pets, you had to feed them and take care of them until they died of natural causes. You lost money and went soft. Jennifer initially wanted to call the farm White Rooster. Kyle demurred. "The prevailing characteristic of these two roosters is not that they are white," he said half-jokingly, "but that they are big and fat." They compromised and called the farm Fat Rooster.

An aerial view of Fat Rooster Farm. The farm is deliberately small, both in scale—20 acres total—and in intention: Jennifer and Kyle don't take on debt, barter for use of equipment, and, as much as possible, look for their own markets to eliminate the middle man.

It is a long-running, if minor, point of contention between Jennifer and Kyle whether the farm's correct address is 354 Morse Road in South Royalton—the mailing address—or, as Kyle maintains, 354 Morse Road in Royalton, a distinction he makes because they are, in fact, a distance from South Royalton, which is located 6 miles south of the farm on Vermont Route 14. Royalton is divided into the villages of North and South Royalton, the latter being the larger of the two, and the center of commerce.

An almanac of the forty-six local towns published in 2002 by the area's largest newspaper, the *Valley News,* records the following data about Royalton, drawn from the 2000 U.S. census. Royalton was incorporated in 1781. Its land area is 41 square miles and its population is roughly 2,603, a figure which includes the student population of Vermont Law School, in South Royalton. (The year-round population is estimated to be around 800.) The population increased by some 300 people from 1990 to 2000. The median family income is nearly $43,000 (the median family income in Hanover, New Hampshire, is, by contrast, nearly $100,000). Some 22 percent of the town's residents are employed in Royalton, and, tellingly, of the top five industries in which its residents work, agriculture appears nowhere on the list.

In 1954, there were some sixty-four farms in Royalton. Fifty years later, there are fifteen, four of which, including Fat Rooster, are organic; most of the rest are dairy farms. If anything, that total number may decline even further in time. The old farmers go out, and their children don't want to continue; the cost of being a small dairy farmer is too great, and the price the milk fetches is too small.

The Upper Valley was settled largely by colonists of English descent who moved up the Connecticut River from Connecticut and Massachusetts, bringing their place names with them: Hartford, Windsor, Springfield, Bethel. In the annals of American Revolutionary War history, Royalton is infamous as the site, in October 1780, of the Royalton Raid, an attack by the British and some 300 Kahnawake Mohawks on the scattered settlements of the White River Valley near Royalton and Tunbridge. In the nineteenth and early twentieth centuries, there was a second wave of immigration from French Canada; many local businesses are still owned by people with names like Courtemanche, Dutille, Boisvert, Levesque, or Hebert.

In the last thirty years, there has been a third wave of immigration, from the cities and the suburbs, which has brought about something of a sea change in the region. Ex-urbs and retirees from New York and Boston and southern New England have trickled in steadily, seeking more affordable housing and property and that more nebulous quantity known as "quality of life," by which is meant proximity to nature and a less frenetic pace within a region that is also within easy striking distance to Montreal, Boston, and New York.

They come seeking the New England beloved of tourist brochures and chambers of commerce: the white-steepled Congregational Church, the town green with the statue of the Union Army soldier leaning on his rifle, the handsome, well-kept eighteenth- and nineteenth-century houses, the pastoral landscape kept open by farming, and the recognizably human scale; what Timothy Dwight, in his 1821 book, *Travels in New England and New York*, idealized as "a place where life may be passed through more pleasantly than in most others, as a place, not where trade compels, but where happiness invites to settle."

All of this is there, to be sure, which makes it all the more disconcerting when the newcomer then encounters an aspect of Vermont that is decidedly not in the brochures: the down-and-out trailers set back in the woods, the tar-paper shacks; the decoy deer (not a real one) standing in the front yard, used for target practice; the junk cars lined up in the backyard; the rusted refrigerators, stoves, and box springs that have been pushed down stream embankments in lieu of paying to take them to the recycling center; and, during hunting season, the pickup trucks that line the dirt roads, gun racks in the rear window, dead bucks strapped to the back.

In many ways it is still a predominantly rural culture, one that does not always brook outside scrutiny or interference, or easily absorb change. There have been clashes between native and flat-lander over the right to hunt, trap, and farm—traditionally considered the domain of the native— as well as access to land for such recreational purposes as hunting, snowmobiling, or all-terrain vehicle

Looking up at the farmhouse and the barn from Morse Road.

(ATV) use. Land that was once open is now posted with NO TRESPASSING signs or alarm system warnings, a collision of values that has given rise to the kind of environmental and class tensions that play out every day of the week throughout the United States: the native asserting rights that he feels have been his for generations, the newcomer maintaining that longevity does not necessarily trump his right to tranquility or security.

South Royalton itself—with its squat, redbrick storefronts and railroad-depot-turned-bank, the laundromat and the pizza joint, the deli and the video store, the bookstore and the copy center, and the abandoned grain mill looming over the railroad tracks—looks less like a preserved-in-amber eighteenth- and nineteenth-century New England hamlet, and more like the working town it is, of a piece with the other mill or lumber towns throughout New England. Its appeal comes not from beauty or grace but from utility, from the fact that it still functions as a small town without an over-lay of preciosity.

The real money lies to the south, in Hanover, New Hampshire, and Norwich, Vermont. Money follows money, which is why those two towns have attracted the well-heeled urban emigrés, and why they have come, in their population and architecture and the kinds of chain-store businesses they attract, to resemble any affluent town in New England or New York State. Money has given them a uniform blandness; they have begun to look like a 1940s or 1950s Hollywood soundstage, a set from *Mr. Blandings Builds His Dream House* or *Christmas in Connecticut,* simulacrums of what people think New England ought to look like.

There is no money on Morse Road or Dearing Road or Clay Wight Road or any of the other roads around the farm. There is, however, considerable beauty. The houses are fewer, the woods more plentiful, the fields more expansive, and the farms, the ones that are still hanging on, cling to the hillsides or sit by the rivers, sending up, in the spring and summer, the pervasive, pungent smell of manure.

The land in this part of the state is a pleasing mixture of mostly wooded, steep hills, numerous intervales, and floodplains on either side of the White and Connecticut Rivers. The geographic term for the eastern part of the state, in which Royalton lies, is the Vermont Piedmont, the land that falls away from the Green Mountains in a gradual, gentle diminution of hills until it reaches the Connecticut River. The rivers of the Piedmont, including the White, drain into the Connecticut or, via Lake Champlain and Lake Memphremagog near the Canadian border, into the Saint Lawrence.

The topography of Vermont—the gouged-out valleys, the rounded hills and the narrow clefts between them—was the result of a mammoth ice sheet that advanced south from Labrador during the last Pleistocene Ice Age, some 10,000 years ago. The glacier scoured the landscape, exposing numerous ledges and leaving behind huge boulders (glacial erratics) strewn in the woods or in the fields, along with great deposits of sands, gravel, and clays. The winter is cold and never-ending; it persists long past the point where you think it should. The spring is a mire of mud and rain and bluster that softens to pink and green come May. The summer is verdant but brief—100 to 130 days: a short growing season, from a farmer's point of view. The autumn is a last, glorious burgeoning of color that draws the "leaf-peepers," a phenomenon on which the local economy is heavily dependent but which predictably draws grumbling—too much traffic, cars that dawdle maddeningly on the back roads, and hordes of camera-toting tourists who take pictures of the trees in your yard.

There is a substantial drop-off from the farmhouse and barn to the fields below, and a second, even steeper drop-off, about 30 feet, from the fields to the lower pasture, down wooded embankments that, in spring and summer, are a tangle of wild leeks, blue cohosh, hellebore, baneberry, and Solomon's seal. There are woods: a rich inventory of American elm, red maple, slippery elm, sugar maple, and basswood.

There are animals and birds, seen and heard, or more often unseen, and the signs that give them away: the dog-like tracks of the coyote and red fox; the dainty, cloven hoofprints of white-tailed deer; the tree trunks that look as if they have been felled and whittled away to a pencil point, that are the work of beavers; the mounds of earth and deeply dug holes in the fields and gardens that indicate the dens of the groundhog, which, if not kept at bay, will eat your entire garden; the telltale scratchy, three-pronged tracks of wild turkey, which travel in large flocks. Red-tailed hawks glide in ever-closing circles over the fields and woods. Reclusive ravens, whose guttural, strangled squawks are heard deep in the woods, or are seen at the edges of fields, dive and roll in the air.

The first mark of spring in March is the return of the red-winged blackbird, with its insistent, prolonged *conk-la-ree.* In summer, bobolinks nest in the fields, you can see the flutter of their black-and-white wings as they rise out of the long grass, and hear their rolling song. There is the iridescent, azure blue of the indigo bunting, and warblers, yellow- and chestnut-sided. The occasional bald eagle will migrate through.

When you reach the lower pasture, where Jennifer and Kyle sometimes put the sheep, it is impossible to see the farm sitting up on the road. The second branch of the White River makes several winding, hairpin turns here—in truth, it looks more like a brook in these narrows—and during migrating season, it's not uncommon to hear cries of wood ducks and mallard ducks and Canada geese down on the water, where they've stopped, or to see them rising steadily into the air as they take wing.

From this vantage point, you can look over to the other side of the river and see the farm belonging to Ella Hyde. It's an orderly white farmhouse with green trim and, in spring and summer, tidy rectangles of plowed field. In late fall and winter, when the leaves are off the trees, you can, driving south on Route 14 by Ella Hyde's, look to your right and see the white farmhouse and the faded gray barn perched on the lip of the rise.

The fields immediately below the farmhouse comprise nearly 11 acres, of which 2 are in cultivation. The remaining 9 acres are used as grazing pasture for the sheep and cows; the house and barn and land surrounding them make up the remaining 5 acres. The fields sit in a shallow bowl of land, so that the effect is of an overturned hat. You stand on the brim of the hat, where the farm and barn are, and look down into the up-ended crown. Below that are the 4 acres of lower pasture by the river. To the northwest are 55 acres of land leased from a neighbor that they use for hay, and across the street, some 35 acres of steep hillside pasture that their neighbors, Charlie and Debbie Morse, let them use for free. In return for allowing Kyle and Jennifer to tap their sugar maples, Charlie and Debbie receive free syrup.

At the very northern edge of the property, tucked up in a corner, is a small pond that Kyle and Jennifer use for watering the livestock. Not far from the pond, set back against trees, abutting the property of their next-door neighbor, are the beehives, four in all, from which Kyle collects honey in the summer.

There is no sense of horizon stretching away here, as there is in the Midwest and West. The perspective is fixed, and finite—hemmed in. You can see your neighbor's property and he can see yours, and you can both see the neighbor on the other side, and everyone knows, roughly, what everyone else is doing. George and Agnes Spaulding live a half mile down the road on Route 14; George is up feeding his heifers—you can see his gangly figure walking across the road to the barn. Chet and Betty Morse live up the hill in a modular home; they owned Morse Farm before passing it on to their sons, Charlie and Bruce, who, in turn, divided it and sold a parcel of it to Jennifer and Kyle. Chet is too ill to get out in the winter, and Betty looks after him, although she has her own physical ailments with which to contend. Charlie, Chet and Betty's son, and his wife Debbie live in a house just below Chet and Betty's. Debbie takes care of Brad a few days a week when Jennifer and Kyle work at their jobs off the farm. In a pinch, when some crisis hits the farm, something that has to be done immediately and by Kyle and Jennifer alone, she will take Brad at no charge. Down the road are two women who moved up from New York and paid a small fortune for a handsome redbrick house with a mansard roof, but who, after a few years, are moving closer to suburban Hanover because this is just a little too remote. Too much rural in the rural life.

The neighbors know if you paint your house and they know if your animals get out, because sometimes the animals end up in their yard, and they wonder whose car that is in the driveway, never seen it before, and they know if you're away, and they know when you're not, and they know when your car has broken down, because you're not driving it anymore, and they know if you've

With burdock in their wool showing their wandering, the sheep and Michael the guard horse are led to another pasture after an escape. The timing wasn't good—Jennifer and Kyle were trying to get the truck loaded to go to the Saturday market.

borrowed a truck from someone else, because they've seen so-and-so in that truck down in town and they know what it looks like.

This is what community means, like it or not, and sometimes you don't. Community means a helping hand and a shared concern over what happens to the land around you, but it can also mean grumbling and tense exchanges and expectations that go unrealized. Farmers aren't rural saints, they're human beings, with all the complexity and ambiguity that implies.

In the spring of 1999, when Jennifer was pregnant, the couple sought the advice of a marriage counselor. The reasons for doing this were, in one sense, obvious—their marriage of nearly four years was under strain—and yet, in the manner of all marriages, complicated by factors particular to them and their life together.

They had bought the farm in July 1998. Jennifer Megyesi had always wanted to farm, ever since she was a kid growing up in East Middlebury, Vermont. Animals were her passion; she had devoted her life, in one way or another, to their care. Not long after she met Kyle Jones in 1992 at graduate school at the University of Massachusetts, Amherst, where they were both working toward their master's degrees in wildlife biology and conservation, she'd told him, "I want to farm," by which she meant farm in Vermont, the place she'd been trying to get back to for the last nine years.

And Kyle had said to her, "I have a farm," by which he meant his family's place out in Circleville, Ohio, in the southern part of the state, not far from Columbus. Here were hundreds of acres of fertile, rolling land on which his parents had raised corn, soybeans, wheat, cattle, hay, pigs, and sheep, and where he had learned to drive a tractor at five years of age. What he didn't tell Jennifer was that he'd borrowed that line (a line which worked, by the way) from the opening of *Out of Africa*—"I had a farm at the foot of the Ngong Hills"—and, more to the point, that farming was something he'd done his level best to get away from, and had not anticipated going back to, until he had met her and fallen in love with her and been persuaded by her vision of a future in which they would farm together.

It all seemed quite romantic—to be in love, to buy a farm, to make concrete the kind of intellectual and ethical and environmental principles by which they conducted themselves—even if, while they looked for the right property to come on the market, they had to live for nine months in a beat-up trailer in East Bethel, with mushrooms growing up through holes in the floor and a furnace that didn't work and an improbable menagerie of animals—a horse, one goat, forty chickens, six sheep, two ducks, and one goose, all belonging to them—housed in an attached shed. These less-than-ideal conditions were only temporary. All it would take was the right place, and their will to make it work.

Given their energies and intelligence and commitment to it, farming seemed like a fairly sound proposition, and one that had appealed to their desire both for autonomy and for community. If they ran a farm, they answered to no one but themselves. Mistakes might be made, and they might stumble, but it was of their choosing; a life that, for all its difficulties, would never bore them. Many people—what Kyle called "9-to-5 drones"—worked at jobs that, over a lifetime, dulled the senses, both mentally and physically. The kind of small farming Jennifer and Kyle envisioned themselves doing was the antithesis, a life in which all the senses were engaged, and which required mental agility and physical endurance nearly every minute of every day.

Jennifer and Kyle had approached acquiring a property with a shrewd tenacity. In the fall of 1997, they had sat down with a map of the area and taking Woodstock, Vermont, as the starting point— where Kyle had just started working that summer as a park resources manager for the Marsh-Billings-Rockefeller National Historical Park, part of the National Park Service—they had drawn a 25-mile radius around Woodstock. Somewhere in there they hoped they would find a farm that was either for sale, or would be for sale soon—a property they could talk its owner into relinquishing, which seemed not unlikely if you drove enough of the back roads and saw all the places that had once been farms, or were now only marginally farms, the farms falling into the ground through neglect and the farms that looked as though only a few more weeks stood between them and a wrecking ball.

Jennifer did just that, road after road, taking down names from mailboxes and, if there was no mailbox or no house number, going to the town clerk's office to track down an owner, and going

Redwing onions on display for sale at the farmers' market. Because of the clay-like consistency of the soil, onions grow well at the farm.

to the local banks to see which properties were in foreclosure. She was working part-time doing wildlife assessment for the U.S. Fish and Wildlife Service, which entailed fieldwork, so she had the time to also scour the area for the kind of property that would be attractive to them.

She and Kyle then sat down and composed a letter that would be mailed to all those property owners whose names she'd gleaned from the mailboxes and the phone book and the town clerks and the banks. It began, cheerfully, "Hello! We are a young, married couple, and have finally made it back to Vermont," and ended by saying, "If you know of any farmland or farms for sale, preferably with old structures, please let us know at the address on the back!" In between was a hopeful sentence about how they wanted to raise "poultry, honey, sheep, market vegetables, and draft horses for market, and raise goats, hogs, and other miscellany for ourselves."

Jennifer mailed out a bunch of these letters, maybe eighty in all, and received, in return, absolutely nothing. No phone calls, no messages, no letters, written in a crabbed, elderly hand, saying, "Why, it so happens that Mother and I have been wanting to sell our precious little piece of heaven, at a greatly reduced fee, to a nice young couple such as yourselves in the hopes that you could start a family there and return the farm to the bountiful glory it once was."

It wasn't until Kyle pulled off the road at a farm to get some maple syrup and chatted with the owner, a man named Melbourne Perley, that they began to get a line on some properties that might be opening up. Melbourne Perley was eighty-nine at the time and he knew the area as well as anyone, maybe better, and he liked to talk. He called Jennifer up and said, tantalizingly, "Come over, I want to tell you where all the widows live." Elderly widows were a good bet, often saddled with farms they couldn't or didn't want to run anymore.

There was the widow Ella Hyde, who lived on Route 14 in Royalton, and there was the widow Medora Dodge, on Slack Hill Road in Royalton, and there was a widow on Chelsea Mountain Road in Randolph and one on a back road to Bethel, to whom Jennifer talked twice. And there was a last place, to which no widow was attached, but rather a family named Morse, who had owned the farm since the late 1930s. There were family tensions there, between two middle-aged brothers, Charlie and Bruce Morse, and a dispute over what should happen to the property, which had been for sale since 1993.

Jennifer went over to see Charlie and his wife, Debbie, who lived directly across the road from the farm. "Yes, it's for sale," Debbie Morse told her, although the way she said it suggested there was some hidden story there. Jennifer then got in touch with Bruce Morse, the younger brother, who didn't live there anymore but down in Massachusetts, and whose property it was legally. When she called, he said, "Oh, you're the people who sent us the postcard."

One point in their favor was that, unlike other people who'd expressed an interest in the property, Kyle and Jennifer not only did *not* want to tear down the farmhouse and replace it with a trailer or a double-wide or a ranch house, but they wanted to resuscitate it, turn it back into a working farm again, which it hadn't been in years. Then followed three or four months of negotiation over the price and the size of the property, with all the usual back-and-forth between the lawyers and the usual moment of impending doom where it looks as though the whole deal will fall through because you can't agree on the terms.

Jennifer had written a letter to Bruce Morse and his wife, Nancy, in which she said that, bottom line, they could only afford to pay $90,000 for the farmhouse and the twenty acres, not the asking price of $120,000, and they weren't going to back down from that. (The price of $90,000 to $100,000 for a house and twenty acres was, Kyle says, "relatively cheap, but expensive for us.") The talk was tough, but inwardly she felt sure they were going to lose the farm, and this depressed her to no end because she'd fallen in love with the place. Then, for whatever reason, Bruce Morse had called back, and said, "Yes, let's do it." The sale was finalized in July 1998, and by the end of July Jennifer was writing down in her farm journal, "Just 4 weeks prior, I had been a tenant in a mobile home (aka trailer) with a lime green carpet and windows that would open if fixed. To me, our newly acquired land was paradise."

It had been much more difficult to get paradise up and running than they had anticipated. Their families thought they were crazy to contemplate abandoning careers that promised advancement and benefits and pensions, for something with a high failure rate and very little margin for error. "I think Dad said, 'What the hell are you going to do *that* for?'" Kyle remembers.

Farming, they were warned, could suck all the money and the life right out of them—quite literally—as Kyle's family could testify. Kyle's maternal grandfather, William Defenbaugh, had died in 1961 in a farming accident when a truck bed he had been working under fell, crushing him. His death had brought Kyle's parents, Alexander "Sandy" Jones and Lois Jones, back from Georgia where they'd been living with their three young sons—Alex, Stewart, and Kyle—to take over the farm's operation. (A fourth child, Stephanie, was born later.) And that wasn't the end of it. Sandy Jones had been badly injured in the early 1980s when a disc harrow ran over him and busted up his knees, and this same farm had ended by bankrupting Kyle's parents in the late 1980s. With the exception of visits home, Kyle had put as much distance between himself and farming as he possibly could. It wasn't just that farming had made a shambles of his parents' finances, but that farming—the haying, the feeding, the working from dawn until nightfall—seemed to be the sum total of their lives.

None of the four Jones kids wanted to farm, and who could blame them?

Yet here he was, some twenty years later, and it seemed just as precarious and anxiety-producing as everyone had predicted it might be. After all, they'd never really done it before. Yes, Kyle had grown up on a farm, and he'd gone through Future Farmers of America in high school and helped with chores, but that wasn't the same thing as owning and running one. Yes, Jennifer had worked on farms and she loved animals, but that also wasn't the same as owning and running one, or caring for livestock. They read books, lots of books: books on living the good life and sustainability and animal husbandry and soil; books by Wendell Berry and Scott and Helen Nearing, whose extended meditations on engaging with the landscape had influenced countless numbers of men and women to migrate back from the cities to the countryside; every possible book related to farming they could get their hands on. But the getting of wisdom was not something that could be gleaned entirely from the books.

Farming, as Kyle observed, is learn as you go. Its lessons were humbling. How could something so pretty and pastoral as a small farm, something that they had both sought out with all due deliberation, seem to thwart them at every turn? It threatened to take all their money and all their time, and it wore them down. The sheer difficulty of making it go began to fray their nerves and sour their idealism, and they began to blame each other for something that had begun to have the unsettling tincture of failure to it.

And now, a year after buying the farm, they were awaiting the birth of their first child, an event that filled them with as much trepidation as it did excitement. They worried about what kind of parents they might be. They worried about the farm. They worried about their marriage. They had almost no money, and tended to have sharp disagreements over how things should be done. Both of them were possessed of a certain degree of stubborn righteousness, and both were unhappy, although if you pressed them they would concede readily that their outlook on things—their sense of how to lead a life that, without being pompous about it, had a moral center—was remarkably similar.

Principles were one thing, but financial reality was another. They'd spent close to $20,000 of their savings to acquire livestock. They did it all themselves and didn't take out loans for equipment that would have saved labor. NO DEBT was their bottom line. Better to struggle than to live, as so many farmers did, under the crushing financial burden that debt imposed. Because they didn't have equipment, they had to borrow it from neighbors, the few who were left in the valley who did farm. Sometimes the borrowed equipment was too old or too broken-down or too rickety to do the job efficiently. Initially, some of the people they'd turned to for assistance were wary, not unfriendly but cautious, because Jennifer and Kyle said they were going to run an organic farm—and in a valley in which many of the farmers were native Vermonters of an insular and conservative stripe, "organic" was synonymous with "hippie." "Organic" meant a certain moral posturing, a zealousness, a lot of starry-eyed, pie-in-the-sky, highfalutin, tree-hugging, nature-loving, pious, finger-wagging rhetoric. No farmer who has been farming for fifty years wants to be lectured by someone who hasn't been doing it for as long as he has, much less by someone who probably hasn't even been born here, and no farmer wants to be told what to do, period.

Jennifer and Kyle weren't like that—they were pragmatic, not didactic—but the farmers around them didn't know that at first and were also skeptical of the barter system that, in some instances, Jennifer and Kyle proposed in lieu of cash payments they simply couldn't afford to make.

You loan us a tractor, and we'll help you come maple sugaring time. You loan us this boar so we can impregnate our sows, and we'll give you one of the piglets. You help us with haying, and we'll look after your animals when you're away, or too sick to do it yourself.

Through a gradual accumulation of trust and goodwill, and the fact that both of them were clearly unafraid of the hard work that farming asks of the people who do it, Jennifer and Kyle won them over. What was funny was that this was the way farmers around here had always done it in the old days, and yet it seemed, in the face of all the changes that had come to farming—the subsidies, the loans, the government programs—radical somehow, to barter goods and services, instead of paying for them or going to a bank for one loan after another.

There were other financial and logistical challenges, one hurdle after another to be surmounted.

There was the property itself, which had fallen into disrepair. It was littered with stuff that had to be removed before Jennifer and Kyle could even begin to undertake the most rudimentary changes on the farm. There were rusted appliances, accumulated junk, and silage bags that had been buried in the fields—all of which, combined with the passage of time, had made of the soil an ineffective, rough instrument, a microcosm that while not completely devoid of life, did not teem with it either. Once that archaeological top layer was gone, they attacked the soil itself. Pastures that hadn't been cut in years, or ever, were now mowed. The soil was fertilized with manure. Lime was spread on the upper pasture, and bonemeal on the lower, river pasture, to rid it of the pervasive, nuisance species of Canada thistle.

The first insurance policy they had on the farm was not specifically a farm policy, but a general house insurance policy. It was canceled when the adjustor came out for an inspection and, claiming the abundance of spiderwebs in the barn were a fire hazard, kicked them off the rolls. Kyle later heard rumors that the same adjustor had, in an excess of zeal, terminated forty other farm policies at the same time.

Jennifer and Kyle didn't have a well, but relied on a spring. The springwater came down to them through a gravity-fed supply line, from a cistern that was up on Morse property; a deeded water right without which Jennifer and Kyle would not have bought the property. Thing was, a spring was only as good as the previous winter's snowpack and the rainfall. Fine in a wet season, trouble in a dry one, disaster during a drought.

The soil had its own idiosyncrasies: because it was composed mostly of heavy clay, it easily retained moisture, which helped to alleviate the ill effects of drought but made it more difficult to handle in a wet year, when it was muddy and clumpy and sodden. Too much rain was almost as bad as too little: plants liked soil that was loose and fine-grained, that would fall through your open fingers like sand, not stick to them like modeling clay.

Then there were the horses, big, beautiful, chestnut Suffolk Punch draft horses with easy, powerful musculature: Hosmer and Ruby, and Ruby's foal, Bobby, that was born on the farm. Jennifer and Kyle had bought them right off the bat, almost before anything else, because they'd had the idea they could use them to plow the fields. If they used draft horses, they wouldn't have to buy or borrow tractors, and it would be gentler and better for the environment—no pollutants from heavy machinery, plentiful deposits of manure—and certainly prettier. But when they tried working with the horses, they found, to their chagrin, that it took too long to plow furrows. They weren't experienced or facile enough to do it quickly, so they gave up, even though they'd spent thousands of dollars to acquire these work animals which they now would not use. (It was sentiment and

Kyle disc harrows a space for a garden.

stubbornness and the hope that they could find some way to use the horses that made Jennifer and Kyle hold onto the horses longer than they should have, as they would later realize.)

All of these, obstacles though they were, did not dissuade Jennifer and Kyle. They were determined to be organic farmers. If they could not farm organically, they would not farm at all. (Time has not swayed them from this point of view, only enforced it.) Both felt strongly that the methods used were much better for land and humans in the long run than the methods favored by non-organic, or conventional, farmers.

Better to have a farm that was a bit haphazard, a bit unpredictable, than what Wendell Berry calls the "monotonous, sterile neatness" of a monocultured farm. Organic meant no pesticides or herbicides or petrochemicals to put undue stress on the land, or to pass on to the people who ate the produce; no reliance on growth hormones or pumped-up animal feeds or antibiotics. Organic meant animals pastured on grass, and animals on grass produced manure in quantity, which was the equivalent of black gold.

If you could take care of the land by rotating crops, by managing grazing, by using the manure from your livestock as fertilizer, by letting sections of it lie fallow a year at a time, the land would, in a sense, take care of itself, unlike land that is leached of its nutrients, its vitality, its elasticity and suppleness, by repeated use of chemical application. That kind of land had to be continually propped up by chemicals in order to keep going, and after a certain point, it couldn't rejuvenate itself anymore. That kind of land would take generations to recover.

Of course, taking care of their land would require intervention and prudent management, just as livestock required intervention and prudent management. Farming is not something that just happens. The simplicity of a small farm—that natural simplicity whose virtues we all extol—is deceptive. A farm is an extraordinarily complex system of interdependent systems and resources—soil, water, air, feed, manure, microbes—and nowhere is that more true than in organic farming, where you have to continually work to keep the soil, water, air, feed, manure, and microbes in an efficacious balance.

And further, to ensure the integrity of the organic chain, from the soil on up, everything—everything!—must be tracked and recorded. A certified organic farmer must, and should be able to, look at any given product and know where and how it started, because the people who buy from you want to know, and being conscientious, you want to know yourself. "We felt it was important to be certified," says Kyle, "as opposed to just saying we were organic." (Until the implementation of the National Organic Standards in 2002, a farm could assert it was organic without certification.) To that end, Jennifer and Kyle had, in 2001, sought certification from the Northeast Organic Farming Association of Vermont (NOFA-VT), an accredited nongovernmental organization that inspects and certifies farms as organic annually, according to a stringent set of standards and quality controls.

Where did the heirloom tomato come from, a NOFA inspector might ask? An organic seed company? The seeds of last year's crop? Or did you swap a Cherokee purple seed for someone else's Omar's Lebanese? Can you be sure, if you're swapping, that the seed is organic? How was the seed kept? Well away from any contaminants? What about the greenhouse in which you started the seeds in February? Where is this greenhouse situated? It's not your greenhouse? You're borrowing it? Whose greenhouse is it? What do they do? They farm, too? What do they farm? How do they farm? What is the earth in the greenhouse like? What was it last used for? Flowers? What kind of flowers? Did the person who grew the flowers use chemical fertilizer? Where was that fertilizer stored? Not near the greenhouse. Are you sure? What is the greenhouse made of? Wood? What kind of wood? Is it treated wood? You don't know? It probably is? That's a problem. Treated wood has arsenic in it. Arsenic is poisonous. Arsenic could transfer somehow to all those thousands of plants you're growing. Seems unlikely, but it might happen. Have to assume it could. Have to correct that problem. If you can't correct that problem in a timely fashion, you could lose your certification.

But cleaving to an organic ideal had other difficulties apart from the rigor it imposed on record and bookkeeping. Conventional farms, whatever their stock-in-trade, tended to rely on the same assumptions of what made a farm work—managerially, financially, scientifically. If you wanted to

The unusually shaped Costolutos Genovese tomatoes—an heirloom variety—catch the attention of potential customers at the farmers' markets.

go into farming, and you chose to operate conventionally, there were numerous models to study, both successful and not. Organic farms tended to be more idiosyncratic; "more seat-of-the-pants," said Kyle. Trial and error. Organically grown produce required greater manual and physical labor, the kind of low-wage, hard physical labor that many Americans don't do anymore, but leave to newly arrived immigrant workers.

In the absence of herbicides, for example, weeds flourish. They must be eradicated before they choke the air and life out of the crop you are trying to grow. To remove them, you get down on your hands and knees in the dirt, often for hours at a time, and pull them out. Or, using a mechanical weed cutter, you might walk slowly down a row of, say, chard or lettuce, and if it was really hot, the sweat might roll into your eyes, and you would end up, hours later, with a back or a neck that ached and streaks of dirt on your body where the sweat had rolled, and hands that were encrusted with sand and dirt, and fingernails that would never really be clean.

And just because that was done, it didn't mean you could rest. Then you would have to go take care of the animals, which posed an entirely different set of responsibilities and problems. You would emerge from the animal stalls with shit on your shoes and maybe hay in your hair or on your clothes and the smell of shit all over you, an almost palpable haze that seemed to have permeated your very skin and which you barely noticed anymore, but everyone else did.

Similarly, in the absence of pesticides, insects proliferate. There is the intimidating 4-inch tomato hornworm, which is the larval stage of a Sphinx moth; the hornworm attacks the tomato's leaves and stalks and sometimes waves its horns at you and also likes to dine on eggplant and peppers. There is the striped Colorado potato beetle that lays its orange eggs, in quantity, on the underside of the potato leaf, where they look oddly like salmon roe; the Colorado potato beetle is never happier than when it is methodically and systematically eating, leaf by leaf, your entire potato crop. There is the flea beetle, as minuscule as its name implies, whose fondness for broccoli and cabbage leads it to make hundreds of pinholes in their leaves; if left unchecked, all that is left of the leaf is the skeleton of its veins, and nothing else.

And when a plant loses its leaves, it dies. Where insects thrive, they, like weeds, must be eradicated. There may be thousands of insects, thousands of eggs, all of which must be removed by hand from every leaf.

And it was just the two of them.

All of these pressures and obligations combined had brought Jennifer and Kyle to a crisis point. "We had no money, and I sat down at the table and screamed," Jennifer says. The marriage counselor heard them out and told them, among other things, that they needed to make a plan, and that it was advisable to do so in a neutral setting. So they went to the closest place they could think of—Village Pizza, two miles south of the farm on Route 14—and tried writing it down, but that got bogged down in disagreements over language and objectives and, more importantly, what kind of farm it would be.

Would it be a farm that they would run as a profit-making enterprise, which was Kyle's position? Or would they just do it out of a love of farming, which had the ring of purity to it but had its impractical side?

It wasn't that Jennifer objected to making money—who would object to that, particularly with a child about to be born?—but that making money was not her primary objective in farming. It sounded corny and impossibly idealistic, but conservation of the land and caring for animals, those were principles to live by. And now this farm would be a place to raise their child, so the stakes had just increased by tenfold.

It wasn't until sometime in the winter of 2000, after their son, Bradford Kipling Jones, was born, that they managed, finally, to get the Fat Rooster manifesto down on paper. It read:

FAT ROOSTER FARM

Jennifer Megyesi and Kyle Jones
354 MORSE ROAD
SOUTH ROYALTON, VT 05068

GOALS:

- *BE NICE TO EACH OTHER*

- To raise and market grass-fed animals profitably, using our own forages as the sole food source to the extent possible, with an emphasis on organic and sustainable agricultural methods.

- To grow/produce meat, vegetables, fruits, honey, syrup, wine and beer, firewood and lumber, and eggs for our own, our immediate family's and our neighbors consumption, and to barter with other local producers for products we can't grow ourselves, thereby allowing us to live without supporting large-scale agriculture and corporate food enterprises.

- To grow and market forages, vegetables, fruits, flowers, wood-turned products and furniture from Vermont woods, on a subscription basis, at farmer's markets, and at wholesale outlets.

- To raise our son using a foundation based upon self-reliance for food, a strong work ethic, a responsibility for and understanding of the importance of land conservation and stewardship, and a respect for his community and the natural world.

Beneath the letterhead and before the lofty enunciation of standards and practices, Kyle had also penned in neatly, *Be nice to each other.*

The summers of 2001 and 2002 were marked in Vermont, as they were throughout the Northeast, by drought—weeks of hot, glaring sun and no rain—and Kyle and Jennifer had to tap into the well belonging to Chet and Betty Morse, a favor that the Morses gladly rendered, but one that made it clear to Kyle and Jennifer they could no longer rely solely on the kindness of neighbors, particularly if the drought continued into 2003.

In the fall of 2002, they applied for financial assistance from the USDA, through a cost-share program run by the Farm Service Agency, and finally dug their own well to the tune of some $10,000, half of which was paid for by the Farm Service Agency. (To secure the remaining $5,000, Jennifer and Kyle finally sold Hosmer, one of the draft horses, as well as drawing on their own savings.) The well enabled them to get water to the livestock much more consistently and quickly, through pressurized water lines. This would limit the damage that would be caused by livestock trampling over the pastures looking for water in the brook or the river, as well as preventing the release of effluents into those water sources.

The well would also enable them to irrigate the crops, rather than rely on the rain or the spring-fed well—a decision that they are leaving for the future. Technology, even something as basic as a well, has its advantages, to be sure, but with it, typically, comes expense. Kyle and Jennifer were not opposed to technology, per se; they're not Luddites, for whom innovation and technology are anathema. But they were fanatical on the question of debt. No loans, no debt; no debt, no bankruptcy. If they wanted something, they had to save for it.

There were, and are, many things to be done on the farm to modernize it, to make it conform to their preferred specifications: Revamping the milk house, where they store frozen meat and fresh produce. Modernizing the Old Barn, so that it can be put to better use. Rewiring the barn as a whole, no small task. Putting in an area for the pigs, so that they can get outside once in a while for sunlight. Fixing the steep driveway that circles the house: in winter it is a vertiginous sheet of ice, and in mud season it resembles the Russian front—ruts so deep and ground so swampy that it looks as though a tank has been practicing regular maneuvers. All of these things, of course, take money, considerable sums of money, which Jennifer and Kyle don't have and won't borrow. They'll get these improvements done when they can afford them, a little at a time.

🐓

Turn in any direction at Fat Rooster Farm, and your eye comes up against hills up to 1,200 feet above sea level, and the woods that have grown over them. This is not untouched wilderness; the trees are not old-growth. Even as recently as thirty years ago, many of these same hills were not forested, but cleared for pasture. The view would have been of oddly bare, crumpled hills and ridges.

But in the last thirty years, nature, which famously abhors a vacuum, has been reasserting itself. In the 213 years that Vermont has been a state, with all the imprints of humans laid upon it, the land, once almost entirely cleared for sheep and dairy, over the course of generations is reverting back to woodlands. The hill farms that used to dot the landscape have gone under and disappeared, the families have moved away. The only evidence that they were once there—the archaeological relics of an older civilization—are the cellar holes and the nineteenth-century stone walls that crisscross hill-tops, the old fruit orchards on top of ridges, their trees twisted and bent by age and weather, and the imposing sugar maples, as tall as churches, that were planted more than 150 years ago to provide shade for sheep and cows in open pasture and that are now almost invisible among the birch, spruce, hemlock, and beech that have grown up around them.

Vermont, although typically thought of as such, has not always been a dairy state. In the 1830s and 1840s, Vermont succumbed to Sheep Fever: a kind of twenty-year economic bubble, based on the amassing of merino sheep, that burst when supply outstripped demand, and hundreds of herds of sheep had to be destroyed as a result. In the same era, word came back to New England that there was more fertile, more easily cultivated land to the west in Ohio, Michigan, Indiana, and Illinois. New Englanders abandoned their farms in droves and eagerly headed west.

Kyle's sense of humor is shown in his sign for pie pumpkins.

Those who stayed eventually converted to dairy farming, which was by 1900 the dominant agricultural base of the state. However, in the years following the Depression—which, nationwide, brought about the largest and most dramatic decline in the number of farms, until the Midwestern farm crisis of the 1980s—and World War II, the number of dairy farms in the state dropped from 10,500 in 1954 to 1,600 in 2000.

As of 2004, there are 6,700 farms in Vermont, and 1,415 are dairy; the 5,285 farms that are not dairy are engaged in raising livestock, growing fruits and vegetables, maple sugaring, and beekeeping. Of the 6,700 working farms in the state, 293 are certified organic by NOFA-Vermont, of which 26 are in Windsor County; the total number of certified organic farms is projected to increase to 391 in 2004.

Recent data released by the Economic Research Service, a research arm of the USDA, shows that of the six New England states, Vermont in 2001 had the highest estimated number of acres in certified organic cultivation—more than 30,000 acres. (California, North Dakota, Minnesota, Wisconsin, Iowa, Montana, and Colorado lead the nation in certified organic cropland; Colorado, Texas, and Montana, in organic pasture and rangeland.)

That statistic may be attributed, in part, to a strong consumer demand in and out of the state for Vermont goods, organic or not; the name "Vermont" lends to its exports an aura of handmade, carefully crafted quality. The state has been careful to promote the kind of land-use policies that encourage landowners to keep property open, or working. Despite the decline in dairy farming, there is an infrastructure that supports agriculture overall. Of equal weight, there is a kind of functioning critical mass: the more organic farmers there are in the state, the more it tends to attract other people who want to be organic farmers.

Overall, however, Windsor County has mirrored the national shift away from agriculture as a livelihood. In 1950 there were 2,092 farms in Windsor County, predominantly dairy farms, totaling more than 346,000 acres, which represented 56 percent of the total land area of the county, more than 600,000 acres. In 2002, there were 697 farms, totaling 89,952 acres, or roughly 15 percent of the total land area. Paradoxically, while the number of farms declined, the average acreage in Windsor

County increased, from 165 acres in 1950 to 297 in 1969, before dropping down to 129 acres in 2002—a shift that may be partially explained by the disappearance of hill farms, where acreage was smaller, and the land of poorer quality, than the farms along the Connecticut or White Rivers.

By any measure, Fat Rooster Farm's 20-acre area (or nearly 100 acres, if you count the land that they lease from neighbors) is small, particularly when you consider that the USDA defines a "small" farm—whether it derives its income primarily from the farm, or in combination with off-farm income—as one whose sales are less than $250,000. The 2002 agricultural census indicates that roughly 92.7 percent of all American farms, according to value of sales, fall into the "small" category. (The National Commission on Small Farms, established in 1997 by then Secretary of Agriculture, Dan Glickman, defined nine out of ten U.S. farms as "small farms.")

The official definition of a farm or ranch is anything where $1,000 or more of agricultural products have been sold. A large family farm is defined as one whose sales are between $250,000 and $499,000; very large as one with sales of $500,000 or more. By contrast, Fat Rooster's gross farm income in 2003 was close to $35,000; its off-farm income in the neighborhood of $26,000; its expenses approximately $25,000.

Because Jennifer and Kyle farm, they are eligible for, and part of, Vermont's Current Use Program (or, Agricultural and Managed Forest Land Use Value Program), in which owners of working farm- or forestlands are assessed a tax rate based on the productive value of their land rather than on the market value. (The program came into being in the 1970s, when Vermont legislators realized that rising property values, and thus rising taxes, were driving many small farmers or foresters out of business, because they could no longer afford to pay the taxes on their land.) They receive, at present, a modest tax break, in the neighborhood of $500 annually.

That Jennifer and Kyle do earn money off the farm—Kyle at the Marsh-Billings-Rockefeller National Historical Park in nearby Woodstock, Jennifer as a technician in a veterinary practice in Bethel—is in keeping with relatively recent research by the USDA that shows the paradigm of a small farm is changing. In the 1990s, after decades of a rural "brain drain" to the cities, as well as the recession and "farm crisis" of the 1980s (in which the United States, primarily in the Midwest, lost thousands of farms to foreclosure and bankruptcy), younger people in their late twenties and early thirties, as well as retirees, began to seek out farming. This was a modest reversal of what had been a steep decline.

These newer farmers, some of whom had never farmed before, have had, of necessity, to be shrewder about making a go of it. They do not rely solely on farm income to make a living. They have learned to target specific markets and to bypass the middleman. They are salespeople and educators as much as they are producers of food, an important point in that the people who buy from them know almost nothing about farming but are curious to learn. On the Open House day that Fat Rooster holds in the summer, visitors are almost wide-eyed when they come face-to-face with the animals and the land and the farm for the first time.

The farmhouse was built in 1872 by Charles and Sarah Woodworth, with the help of their daughter, Juniata Flora Woodworth (commonly known as Junie), on land that Charles Woodworth had bought from one Benjamin Cozzens in 1853. Until the building of the farmhouse, the Woodworths had lived in a log cabin across the road from the present farmhouse. Junie Woodworth was born in 1854, the last of five children, and raised on the same property, and it is reasonable to assume, although there is no specific record, that she moved into the house with her parents and siblings. Her father is said to have raised prize cattle and Morgan horses, the oldest extant breed of American horse, first sired in Vermont.

She continued to live there after her parents' death—Charles in 1887, Sally in 1905—and it became her primary residence. She wed in 1897, at the relatively late age of forty-three, a man named Ruel Mills, about whom little is known except that he was born in nearby Bethel and that he lived in the farmhouse with Junie after their marriage. They had no children. She was a widow when she died in August 1938 at the age of eighty-four from nephritis (inflammation of the kidneys).

In July 1938, nearly three weeks before her death, Junie deeded the house and some 100 acres to her great-great-nephew, Henry "Chet" Morse Jr., for the munificent sum of a dollar, with the following stipulation: that Morse and his bride-to-be, Betty Gray, who had already moved in as a caretaker with Junie in the spring of 1938, "shall well and truly provide me with suitable board and domestic care during the remainder of my life, in my said home," and that "Henry C. Morse Jr.... carry on the premises in a husbandlike manner. . . ."

Junie had been born and raised and lived there, and she intended to die there, says Betty Morse. Because Betty Gray was only sixteen then, and not of legal age, she was not mentioned in the will, even though she and Chet were already engaged to be married. When Betty moved in with Junie Mills, who was already quite ill ("She took to bed and didn't get up," recalls Betty Morse), the house was much as it had been when it was built in 1872. There was no electricity, no running water, no toilet, no plumbing—just a two-hole latrine outside, rigged so that it was flushed by rainwater. (Junie Mills kept a chamber pot in her room.) The only heat in the house was a three-legged stove in the kitchen.

To wash your face and hands, you had to go outside, at the back of the house, where water ran into a cement tank; a tree stump had been placed beside the tank, and you put your washbasin on the stump. The house's sole concession to modernity was a wall telephone. One of the first things that Chet and Betty Morse did, upon taking possession, was to install running water and a sink, and dig a sewer system. Three years later, they had saved enough money to put in a bathroom, with a toilet and a bathtub they bought from Sears & Roebuck.

The farmhouse still has the narrow lines and steep pitch of a nineteenth-century building, the kind of architecture that speaks to the pragmatism of the people who built it. It is a house just big enough to live in, and no more. The interior is as narrow as the exterior, even with the modernization of fixtures and wiring and some remodeling. It has the cramped feel of an old farmhouse: low ceilings, narrow, steep stairs, and rooms on the second floor that accommodate only a few items of furniture, all of a proportion to make you think that human beings were simply smaller in size, or that their needs were more modest.

The house and barn as seen from above. Jennifer and Kyle fix up the farm buildings whenever they can.

The interior is still modest, albeit comfortable. The furniture is utilitarian, the kind you would expect to see in a college dorm or in a first apartment, found objects and secondhand tables with the occasional, artisanal piece made by Kyle in his wood shop. A woodstove in the front room radiates heat. There are nineteenth-century lithographs of various bird species on the walls, a forest of plants that Jennifer has maintained for the past seven years, and one aquarium. Brad's toys and books are piled into a corner, near a chair. A small coffee table is strewn with books on farming, and magazines: *Smithsonian, Farming,* flyers from various national or local organic organizations. There is a television, a CD player, a stereo system, and a Sony turntable which Kyle has hung on to, rather than throw away his collection of records. The ceiling near the kitchen is hung with drying flowers, strings of dried chili peppers, and garlic braids. The pantry shelves are lined with home-canned beans, beets, pickles. A small army of boots, usually mud-encrusted, stands in formation by the front door.

It is often hard to tell where the house ends and the farm proper begins, as the traffic of animals and humans between barn and house is fluid and constant. Either there are humans in the barn, or there are animals in the house. It is not unusual to walk in and see ailing lambs in a cardboard box, or the occasional animal or bird, temporarily brought home from Jennifer's job at Country Animal Hospital, to which she tenders care. There is a birdcage containing a maimed chicken, House Hen, that was wounded by an owl in an attack and consequently received indoor status. When the chicks and ducklings come through the mail, they are kept in the bathtub in the first-floor bathroom until they are old and strong enough to go into the barn. (There is a shower, for human use, on the second floor.) Down in the cellar are buckets containing hundreds of fresh eggs that will need to be washed clean of dirt and hay and excrement before they can be sold to the public.

After Chet and Betty Morse took over the farm upon their marriage in September 1938, they had the idea to run it as a dairy farm, starting out with six heifers given them by Chet's father, Henry Morse. There was no barn attached to the house, only an old shed, so they added on to that shed. Around 1942 they finally built a brand-new barn, which stood until 1961, when they began building the barn that Kyle and Jennifer use today. The barn was completed in 1968 when the back end of the barn—what Jennifer calls the Old Barn, although it was the last addition—was grafted onto the structure.

The 100 acres deeded Chet and Betty by Junie Mills originally stretched up the hillside behind the houses where Chet and Betty Morse, and Charlie and Debbie Morse now live and down Morse Road, toward what is called Four Corners, where four dirt roads come together. In 1976, Chet and Betty's two sons, Charlie and Bruce, took over the farm, because doctors had advised Chet, who had had two hip replacements—a common complaint among dairy farmers—that farming and he would be better off parted. The property was finally deeded in 1978 to the two sons.

By the time Kyle and Jennifer came along, in the fall of 1997, it had been farmed fitfully and then rented out as a house. It was clearly not what it had once been. "The farm was run-down, the fields were overgrown, we could see a lot of brush, it was weedy," says Kyle. "A single strand of electric fence doesn't hold in much, except maybe horses."

They were not impressed. It looked bleak and too far gone, although Kyle avers that fall and winter are a good time to look at land because you see it as it really is. Its skeleton is revealed, the spine of the hills and rocks are laid bare, and if you are going to live in a place like this, the time to see it is not at its most lush but at its most barren, because that is how you are going to experience it from November until May.

After seeing the farm again in the spring, when the landscape had the look of what people around here call "green lace," a delicate, tentative filigree of leaves slowly coming into bud across the valleys and hillsides, they reconsidered. "I remember so clearly walking across the fields in May," Jennifer recalled. "We saw wild turkeys, trillium, Dutchman's breeches, and flowers everywhere, and I just burst into tears." The wild leeks were abundant on the embankment leading down to the river; an expanse of green amongst brown and gray. Jennifer had wanted to find a property that had wild leeks, and here they were. It seemed to her to be a harbinger, an augur that this was the place they were meant to be. Here would be a life.

THE FAMILY

The winter of 2003 was one of the coldest and clearest in recent memory. It was the kind of weather in which people disappeared behind hats and scarves and developed a permanent hunch to their shoulders. Old-timers trotted out folk sayings like, "Clear as a bell, cold as hell," or, "When the days do lengthen, the cold doth strengthen," with dark predictions about the spring to come.

Kyle shovels a path to the barn after a storm brought two feet of snow to the area.

Because the two previous winters had been relatively mild with relatively little snowfall, which had contributed to drought conditions in the summer of 2002, the sudden plunge into zero and subzero temperatures in January 2003 came as a rude shock. The thermometer readings dropped steadily: 8 degrees Fahrenheit on January 18, 15 degrees on January 19, a balmy 21 degrees the next day, then down to 8 degrees on January 21, 6 degrees on January 22, 5 degrees on January 23. And these were the highs. At night the temperatures went down to zero, 5 below zero, 10 below, 15, 20, maybe even 30 below zero.

That kind of cold and deep snow would thin the deer herds and the coyotes as no hunting season could, through starvation or cold, or both. Old-timers snorted at the whiners, the weaklings: this is a real Vermont winter; this is the way it used to be when there were 5-foot snowdrifts against the house and you milked the cows in below-zero temperatures, and the milk froze in the pails. The last two mild winters without snow were the aberration; this is the way it is supposed to be.

Entire conversations in northern New England may consist of nothing but talk about weather, which is not some weak conversational gambit born of desperation or lack of imagination, but a matter of the keenest concern and interest. Snow, cold, rain, heat, drought, flood. Everything rises and falls on the weather: your animals, your crops, your machinery, your hay. Farmers know to cast a look at the sky, to see which way the wind is blowing. Is it the dull, cold metal of a winter sky before it snows? The pink sky at dusk that foretells clear weather the next day? The low, dark, scudding clouds that presage rain, or the high cumulus clouds that signify fair skies?

"Birds flying low, expect rain and a blow," the old-timers might say. "A cow's tail to the west is weather coming at its best; a cow's tail to the east is weather coming at its least." "If the moon rises with a halo round, soon we'll tread on deluged ground." Folklore, not science, but the kind of folklore built on generations of empirical observation of New England's fickle weather, of which Mark Twain famously observed that, "There is a sumptuous variety about the New England weather that compels the stranger's admiration—and regret. The weather is always doing something there; always getting up new designs and trying them on the people to see how they will go."

Breaking up an iced-over water dish with his boot, Kyle clears out the pan for fresh water for the chickens. With nighttime temperatures dipping to 20 degrees below, frozen water pipes that can burst are a big concern. Either he or Jennifer check on the barn a couple of times a night to be sure all's well.

Two feet of snow had fallen at the end of December 2002, and was now frozen in place. At night, the farmhouse shook in the winds that came in from the northwest. The sky was blue, the sun shone day after day. It looked as if it should and might be warm, given the dazzling light, but it was not. There was a pristine, if frigid, clarity to everything; the outline of the hills and trees and rocks was absolutely precise and crystalline, finely shaded black and white, as if etched by Dürer. In the bare, gnarled branches of trees, the abandoned birds' nests of summer were now plainly seen. Smoke drifted upward from chimneys the length of the valley. Animal tracks—deer, raccoon, coyote, squirrel—left crisp punctuation marks in the snow, looping across fields or up one side of a hill and down the other. Wraiths of mist hung low over the rivers. Ice choked the streams. At night, in that clean, arctic air, the stars seemed particularly brilliant, particularly close.

January is usually marked by a thaw, but there was none. Nor was there a thaw in February: 10 degrees on February 13, 13 degrees on February 14, 5 degrees on February 15, inching up 1 degree on February 16. And it continued into late February and early March, with cold days and cold nights, by which point even the old-timers were crying Uncle, and worrying about whether it would warm up enough—warm days, cool nights being the optimum condition—so that the sap in the sugar maples would give them a good run. The world outside looked as though it were in a state of suspended animation. Nothing seemed to move. No birds, no animals, no people.

Everyone complained, everyone was miserable. Brad, who had just turned three in early January, hated to go out in it, but he hated the confinement inside just as much. All the big animals were kept in the barn in such extreme cold, not because they would necessarily suffer outside during the day, but because their combined body heat kept the barn at a stable and tolerable temperature; when it was zero or below outside, the interior of the barn during the day hovered at 31 or 30 degrees.

There was something stoic and graceful about the cows and sheep in their passive acceptance of their lot, in their docile bulk, their pungent animal smell, and in the sounds of their collective breathing and the rustling as they worked their jaws and teeth and lips methodically around the dried stalks of hay that they were fed twice daily. The chickens looked unhappier, more woebegone, and often their combs suffered frostbite. They ate eagerly and noisily, lining up on either side of the long food trough in the henhouse as if they were in a high school cafeteria; the sound of their hundred beaks simultaneously hitting the trough was like a heavy rain on a tin roof. All the water in the hens' shallow water dishes froze during the prolonged cold snap, and Kyle had to kick at them to dislodge the crust of ice that had formed overnight. He had put up protective layers of plastic on all the barn windows, which were old and drafty, and that afforded some insulation, but not much—from coldest to colder.

On a day at the end of January, on which the thermometer struggled to rise above 5 degrees, Jennifer stomped around in the barn, nose red from the cold, yelling things like "I hate Vermont" or "This sucks" as she threw hay to the sheep and mucked out the pig stalls that were, even in the chill, ripe with ammonia. Kyle, his breath clearly visible, was hammering out a pen for the sow named Old Spot—so called from her breed name, Gloucestershire Old Spot, an old and rare breed—that was due to farrow (the farmer's word for a sow giving birth) in ten days, at the end of January. The sow named Cookie had already farrowed a litter of fourteen piglets, on Christmas Eve. The sow named White Pig was due to farrow not long after Old Spot, by February 12.

Both Jennifer and Kyle looked like airplane mechanics or sanitation workers, dressed from head to toe in heavy-duty Carhartt coveralls, stiff gloves on their hands, faces pink from the cold and the exertion. Apart from building the farrowing pen for Old Spot and White Pig, Jennifer and Kyle would also move the sows around from pen to pen, a dodgy, potentially dangerous operation, given the sows' size and formidable teeth and notorious intransigence.

There were three wooden pens, with gates on the inside of each that opened into the succeeding pen. Jennifer would push one sow from one pen into the next, slapping it gingerly on its back, and then getting out ahead of it with the most reliable bribe—a grain bucket that the grunting, snuffling sows predictably followed—while Kyle waited on the gate.

Sometimes the sow didn't want to move and made her feelings known with an ear-shattering, terrifying noise that couldn't, in truth, really be called a squeal. "Squeal" sounds too small, too manageable, too piglet-ish. The sound was somewhere between the ferocious, high-pitched whine of a 747 as it touches down on a runway, its engines then thrown into reverse, and the brakes of a Mack truck applied suddenly and with force, thereby sending the truck into an interminable skid. The sheer intimidation factor of a screaming, protesting, 600-pound pig cannot be overestimated.

"Notice I have to get in with her," said Jennifer jokingly, if pointedly. "Kyle is scared of her."

Kyle gave her an edgy, I-can't-believe-you're-saying-this look. There was more back-and-forth between them, a conversation involving the potential acquisition of more animals, with Jennifer typically lobbying for more, and Kyle for less.

"I want to raise rabbits," said Jennifer. "Kyle wants to raise guinea hens. It's the perfect standoff."

"Not really. Guinea hens make sense," Kyle retorted.

Cookie's piglets had begun dying in the two to three weeks after their birth: three from pneumonia; three from anemia. They would be found in the morning, stiff, pinched-looking, eyes closed. The reason seemed evident: the cold.

The piglets had slipped out of their mother's womb and into the world, slick with blood and afterbirth. She had expelled each piglet with a gusty heave of her flanks, an almost noiseless shudder that ran along her body. She lay on her side, eyes closed, resigned, the rippling along her skin the only sign of the contractions she was experiencing. Some of the piglets were the rosy pink of a newborn infant, and others were pink with black spots. Jennifer had wiped the piglets clean and then cut off their umbilical cords, putting iodine on their navels as disinfectant. They lay quietly on the hay, stunned by their abrupt entrance into life.

From there, she thrust them immediately under a heat lamp. It took them a few moments to wobble up, to move shakily, to accustom themselves to the new surroundings. It might be as long as four or five hours before they could nurse; a sow was, theoretically, expected to farrow within two hours, start to finish, at which point she could begin to nurse. Piglets could go as long as five hours after birth without nursing.

Cookie was an unusually amiable creature. She liked to be scratched behind her ears, like a dog, and if you patted her bristly, short hair, a cloud of dust arising, she did not protest or flinch. She was also a good mother, by Jennifer's estimation. She paid fairly close attention to her piglets and made the kind of guttural, grunting noises that encouraged them to nurse, and she was patient with them, unlike the much-despised White Pig, who had once torn apart two of her piglets with her teeth and was likely to turn in a fury on her brood for no other reason than they seemed to irritate her.

White Pig was something of a question. Would she stay or go? She was mean and she was unpredictable and she fought all the time with Cookie and Old Spot, muscling them out of her way to get to the food. Even her little eyes had something sullen about them, a "What's in it for me?" look. (Cookie's expression was somehow friendly and inquisitive, while Old Spot was placid to the point of comatose.)

But, and here was the rub, White Pig was a Yorkshire Landrace—the premier pork pig—and she had what they call good conformation. Her lines and proportions were right, as were those of her piglets, and they were fine, fat pigs when they reached maturity, with fat haunches and meaty shoulders that would almost certainly fetch a good sum in the specialty food markets of New York or Boston, where they would end up as suckling pig or, once they had grown up to fuller size, prosciutto or ham.

Like White Pig and Old Spot, Cookie sometimes crushed her piglets; a certain percentage of piglets were expected to die in each litter from the accidental rolling over of a 600-pound sow onto a piglet that might be anywhere from 5 to 7 pounds. Mortality rates from this sort of accident were part of a farmer's arithmetic; the cold of a New England winter was also part of the equation. But not *this* cold, not cold that was so tenacious, that gave no sign of loosening its grip. A day or two of cold, yes, not weeks of it, and it had taken its toll. Even under the heat lamp that had been hung over the pen for them, the piglets shivered. They piled onto each other for warmth so that they looked like a little squirming mountain, and they scrabbled frantically, with more urgent squeals than usual, for room at their mother's nipples; one piglet to a teat. Once chosen, that teat belonged to that piglet until it had weaned.

Seven of sow Cookie's ten piglets nurse minutes after they were born. The farm sells piglets at about two months old as suckling pigs to New York and Boston restaurants, as freezer pigs at six months old, prosciutto pigs at nine months old, and also to people in the local community to raise for themselves.

Part of the reason for the higher mortality rates was that the sows had been bred in the fall, to take advantage of the Easter markets for suckling pigs in New York and Boston. Jennifer and Kyle had no boar of their own, but, at the appropriate time, rented boars from various farmers, including the butcher, Mark Durkee.

Durkee's boar, who had impregnated Cookie and Old Spot and White Pig in the fall of 2002, was a traveling boar; "Rent-a-boar," Jennifer called him. He went from farm to farm as a sort of porcine gigolo; an operation that involved sticking him in with the sows when they came into heat and hoping for the best. Some boars took to the job with gusto, exhausting and irritating the sows with their attentions; others seemed unsure of themselves; and the occasional boar did nothing at all except eat and sleep, which usually doomed him to an early death. When animals don't produce, they are killed.

Pig sex is noisome, messy, unlovely, and malodorous: "like chlorine," says Jennifer, wrinkling her nose. If the sows become impregnated, their gestation period is three months, three weeks, and three days; if they are bred in the fall, as Jennifer and Kyle had done, the piglets could be born into conditions that might well prove inhospitable, even fatal. Which is what had happened in the winter of 2003.

Another farmer, who had had some fifty piglets, had lost nearly all of them, despite using propane heaters to keep them warm. The conditions were just not good; the cold, the damp, the bacteria breeding in the manure in the stalls, all together affected the lining of the pigs' lungs. In early February, a vet came by Fat Rooster Farm to perform a necropsy on a piglet, dead three days and beginning to smell. After cutting into the pig and examining its intestines and looking at the prevalence of fluid within the body cavity, his conclusion was septicemia, brought on by bacteria.

Eat Rooster was not an industrial operation, by any means—no prefab farrowing pens so small that the sow has not, literally, one inch to move in, no fans to ventilate, no heaters. This was the old school. Now Old Spot was due to farrow soon, and Jennifer and Kyle would have to redouble their efforts to try to keep the piglets alive. To reduce bacterial infection and to minimize the cold moisture that would serve as a breeding ground for the kinds of germs that would lead to pneumonia or septicemia, the stalls would have to be cleaned every day.

Antibiotics, under the standards applied by NOFA, are permitted under certain circumstances—animals aren't allowed to suffer unnecessarily—and these qualified, but Jennifer didn't intend to use them because she felt the problem was one of husbandry, not medication. If the pens could be kept dry and clean and as warm as the weather allowed, the piglets would be at less risk. It would be another three weeks, at least, before it warmed up enough to open the windows and improve air circulation.

Jennifer and Kyle's goals for 2003 were relatively clear: Grow, but within reason, prudently and cautiously enough that they didn't lose too much money. They had both seen other operations that expanded too far, too fast, and then had to retrench or go under. From the beginning, Jennifer and Kyle had divided the farm into two operations, the livestock and the produce, each of them bearing enormous responsibility and demanding, it almost went without saying, hard work. By mutual agreement, animal husbandry was Jennifer's province, not Kyle's. Kyle attended to the farm's infrastructure—repairs, rewiring, tearing down, building up—and, in the spring and summer, did the plowing and haying and baling. This wasn't set in stone; they each did what was needed at any given time, and both of them worked together on any number of jobs or problems, from repairing fence to harvesting to chasing animals if they got loose.

There were in the barn, at the beginning of the year, fifty-one ewes and two rams, seven cows, a hundred laying hens, five pigs, and the two horses. The livestock accounted for 75 percent of the farm's operation, and it was what distinguished Fat Rooster Farm from the other organic farms in the immediate area, which traded primarily in fresh produce. Jennifer and Kyle sold their meat and poultry in four ways: through the Community Supported Agriculture (CSA) program; at farmers' markets; through an occasional sale directly on the farm; and through Vermont Quality Meats, a cooperative partially owned by its suppliers, which sells both conventionally raised and organic meat to well-known restaurants in Boston and New York.

Because of publicity over the safety of the U.S. meat supply, with the specter of mad cow disease, or bovine spongiform encephalopathy (BSE), hovering over American agriculture—the first confirmed case of mad cow disease in the U.S. would be reported in December 2003, in Washington State—the market for organically raised meat, at least in this small corner, appeared to be on the rise.

The flickers of interest could certainly not be called a stampede away from conventionally raised meat to organic, particularly given the price (two to three times as much as conventionally raised). But the publicity given BSE, in conjunction with the use by some farmers, and corporate farms, of growth hormones and antibiotics—as well as questions over treatment of American livestock prior to, and at, slaughter—indicated a segment of the American public had begun to believe that their faith in the integrity of the U.S. meat supply, and the management of it, had perhaps bordered on complacency.

While adding livestock to the farm sounded like a reasonable thing to do if Jennifer and Kyle were interested in making more money, it had inherent challenges. Every aspect of a farm requires managing and hand-holding, but this is more true of livestock. If you lost some lettuce or corn or pumpkins, well, it was unfortunate, but you were not out hundreds or thousands of dollars as you might be if you lost livestock. A non-farmer might have the sense that animals, if left to their own devices, would thrive in the same way that vegetables thrive if they get enough water and sun and nutrients. They're born, they live, they die. In fact, the care and feeding of livestock is remarkably time-consuming and worrying, particularly during birthing. Animals get sick and die, and you have to ask why, and fix it if you can. They are not wild animals, but neither are they pets. Apart from their all-consuming interest in when and where they were going to be fed, most of the birds and animals at Fat Rooster Farm shied away from humans. They functioned as herds or flocks, with an instinctive distrust of humans and what they might do, and yet they were almost utterly dependent on humans for their care and well-being. They occupied a middle ground at Fat Rooster Farm between pet and commodity. There were runts who, because of their size and fragility, were expected to die, but lived long enough to become pets. There were cows who did not become pregnant when they were supposed

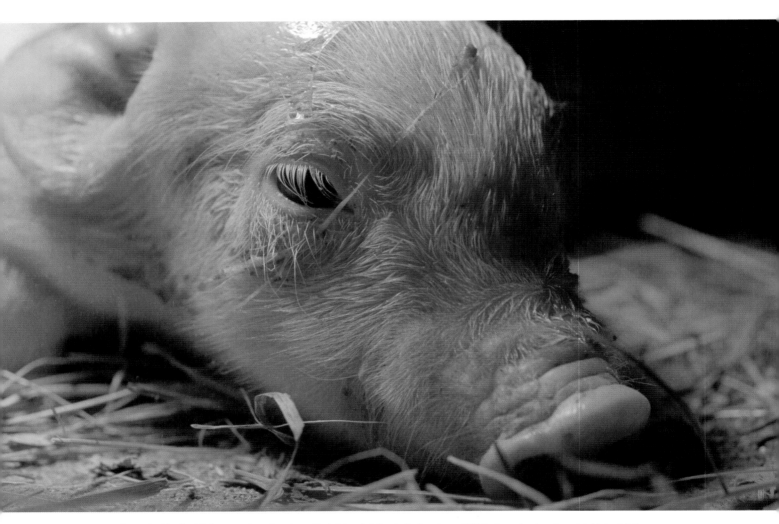

Coated with afterbirth, a newborn Landrace Yorkshire piglet gets its bearings. Within five minutes, the piglet has dried off and is walking over to nurse from its mother.

to, and despite this, were kept on, rather than being sold or shipped to slaughter. There were animals or birds born with deformities that rendered them weaker and more helpless, and despite tough talk about killing them and putting them out of their misery, Jennifer or Kyle let them live.

"Q-Tip was born in September, out of season, a weak little lamb. Was that a business-driven decision?" Jennifer observed. "Our heifer, Brown Sugar, was born during sugaring. There's a hen named Chickie When The Egg Grows. Like that chicken's ever going to leave the farm! I can pick her out of ninety-seven. It's emotionally driven thinking."

Kyle and Jennifer were mulling over increasing the size of the sheep flock, but did not know by how much. The books told them, go from fifty to seventy-five or one hundred sheep, but not from fifty to, say, sixty. If you are going to expand, and spend the extra money and time on the feeding and care of the sheep, it is just as cost-effective to expand by a greater number than a smaller one.

The lambs were profitable. Customers at the Norwich Farmers' Market—Jennifer and Kyle's biggest, most profitable market by far—who could generally afford to pay higher prices, went crazy for baby lamb that was locally and organically grown, and would pay top dollar. Even if the market fell, the margins were such that they would continue to make a good profit on them. As it was currently set up, however, there was only so much space in the barn. The pens could comfortably hold around sixty sheep, but not seventy-five. Any expansion of that scope would have to wait until they could build new pens in the Old Barn, which stood largely unused.

The issue of cattle was ongoing. Kyle was pushing to get out of it altogether because he didn't think they had the fencing to keep the cows corralled. They had a tendency to blow through the fences, and then he ended up chasing them from one pasture to the next. Rounding up a herd of runaway cows, even a small herd, eats away time and money; they break this way and that, stopping just close enough to you to taunt you with their nearness, staring at you unblinkingly as if to say, "Come get me, I dare you." And just when you think you have *that* one in hand, another one gets away.

"We don't have enough fence to keep them in, not enough profit," Kyle would say at the end of 2003, when they were still discussing whether to keep them. "This summer, four times in one day, I had to chase them up on that hill!" His voice rose and, from his seat at the table, he twisted around and pointed out the window, at the steep pasture behind Charlie and Debbie's house.

"Three times," Jennifer corrected him. "I did it the last time." They looked at each other mulishly. The matter remained unresolved.

The word "fence" may mislead. It refers not to a picket fence, but to the high tensile wire in place around the farm's perimeter, and also to the portable polywire netting and electronetting that sections off pasture being grazed. To keep the livestock in place, the fences are electrified so that if cows and sheep blunder into them, they receive a good shock, not enough to incapacitate but enough to give them the idea they shouldn't make a habit of it. The electric fences were not always on, though, and sometimes the animals just didn't care. If the cows and sheep—they were often pastured together—were restless for fresher grass, they could easily plow through fencing.

You could see them lining up on one side of fence staring longingly and impatiently at the long, verdant grass just centimeters away on the other side of the fence, making the kind of noises—a higher-pitched, persistent bellowing and baaah-ing—that put you on notice that if you didn't give them what they wanted, they would take it on their own. "It's a sound that I cannot describe," Jennifer observes, "a sound that says 'Good-bye, I'm leaving.' And they leave."

Jennifer, to Kyle's frustration, was resisting getting rid of the cows. For one thing, she liked them, and had a pet cow named Tildy Ann of which she was very fond. This cow had originally been called Fern, but Melbourne Perley, the farmer who had alerted them to the fact that the old Morse farm was for sale, had said, somewhat derisively, that no self-respecting cow should be called Fern, and he had renamed her Matilda Ann. For another thing, the cows made good beef, and, despite his protestations that he wanted the cows gone, Kyle liked to eat beef. He wouldn't eat lamb, which may have had something to do with his Midwestern upbringing, in which beef was king and every other meat suspect. (Or it may have had something to do with a vein of sentiment for the

With help from Brad, Jennifer installs electronet fencing in a pasture near the barn for the new lambs. It was the first time the lambs were out of the barn—the lambs and their mothers are all put out at once to get them used to the electric fence.

sheep.) In addition, Jennifer pointed out that the cows provided much-needed body heat in the barn during the winter months.

They didn't have a bull, and they didn't want one: hard to move, hard to load, potentially dangerous, and expensive. Instead, they were trying artificial insemination with their cows. When the cows calved, they would raise them until they reached the stage where they would be slaughtered for veal or baby beef. This was veal raised on mother's milk and, for a little while, on grass; not veal penned in an impossibly small space.

The laying hens were fairly constant; about 100, barring predation or accident, and there was no reason to change the number.

The previous year, 2002, they had had in the neighborhood of 500 meat birds, and that had been a disaster. "We took such a bath on them," Jennifer recalled, shaking her head. They didn't have the facilities at the farm to slaughter that many birds, and they paid a man to come do it for them. He had a truck with a trailer designed specifically for slaughtering birds; what, in the business, is called a mobile slaughter unit. The chickens had arrived as chicks in July, and the idea was to kill them in September, when they were eight to nine weeks old, and had reached about four pounds. But he had problems, his truck broke down, and he didn't come until the end of October, by which time they dressed at 12 to 13 pounds, and were, said Jennifer, "tough as nails." And more expensive: "Who wants to pay $25 for a chicken?" Jennifer said rhetorically.

Of the 500 chickens, they had sold 200, but were left with 300 that they couldn't get rid of. That had stung. They would reduce their meat-bird operation to 100 birds for the coming year as a result, and they would have to find another way to slaughter.

They wanted to increase the number of ducks, from the twenty-five they'd raised and killed in 2002, to fifty, because ducks could fetch $4 per pound at market, or $20 per duck.

They hoped to add a few pigs to their current crop of three sows and two pigs being raised for prosciutto. The number of sows was more or less stable—three—as was the number of pigs raised for ham—anywhere from two to six. The pigs were the only animals on the farm not to be raised organically. They were the equivalent of a recycling center or, more accurately, a sludge processing plant. They ate everything they could get their snouts on: vegetable scraps from markets, food scraps from the farm kitchen, milk from neighboring dairy farms; some of it organic, some of it not (even the smallest particle of nonorganic food fed an animal whose diet is otherwise certified organic, renders that animal nonorganic)—and turned it into high-quality manure for the farm. Because they were not considered organic, they were not fed the more expensive organic grain.

The pigs would be sold through Vermont Quality Meats to restaurants in New York or Boston, or bartered with friends and neighbors, a move that wasn't necessarily lucrative but garnered goodwill. If you thought of the farm as the center of a spider's web, then their standing within the community expanded in concentric circles, a delicate network of favors and services and goods exchanged that then brought in as much as they gave out.

After the animals came the question of crop production.

They planned to plow two more acres this year than last, which would triple their vegetable production. For the first time, they would grow flowers, for no other reason than that Jennifer liked flowers, and because she thought she could sell them at farmers' markets—a proposition of which Kyle was dubious. Kyle, on the other hand, wanted to try growing shiitake mushrooms, something he had never done before, and for which he thought he could get good money—a proposition of which Jennifer was equally dubious.

This year, they had arranged with Stewart Skrill, a sheep farmer in South Randolph to lease one of his two, 96-foot greenhouses, standing empty, that he had originally used to grow flowers for the market in Boston. (The greenhouse they had leased in 2002 would no longer be available to them, because the owner had decided to take it back for her own business.) In return, they would pay Skrill a nominal rent of $10 per month, foot the fuel bill themselves, and grow plants in the greenhouse for his wife. This greenhouse was of far greater capacity than the one they had rented the

While Brad plays on a table, Kyle and Jennifer mark sticks that label seed flats in a greenhouse they rent from Stewart and Karen Skrill in South Randolph, Vermont. Not a fan of long Vermont winters, Kyle especially enjoys working in the warmth of the greenhouse.

previous year, and could easily hold the thousands of seedlings that would emerge and grow during February and March and April, before being transplanted into the ground at the farm at the end of May when the danger of frost had passed.

Between the seeds they'd exchanged with other farmers, the seeds they'd saved from the year before, and the seeds they had bought from certified organic growers, as well as other sources, they would be planting 256 varieties of vegetable, herb, or flower come late spring. It had cost them roughly $1,000 to buy seeds, but they had undoubtedly saved at least $1,000 more through the preserving and bartering of seeds.

They wanted to expand markets, from Norwich Farmers' Market to the Lebanon, New Hampshire, market, which would be in its first year and a complete unknown. They would continue to sell produce through food cooperatives in South Royalton and Randolph, Vermont.

This year, Jennifer and Kyle had increased the number of CSA shares: nineteen families would sign up for the program. (Two would drop out over the summer.) The previous year, the first in which they had implemented the program, there had been ten families. For $600, paid in advance in February, each shareholder could expect to receive a certain amount of food weekly from the end of May through October, with the bulk of food coming during the summer months, when the vegetable production was at its peak.

Shareholders were also assured of a weekly supply of fresh eggs, a certain amount of maple syrup, an allotment of chicken, duck, and lamb, all raised on the farm, and this year, flowers. People who did not want, or could not afford, to pay the full $600 could pay a half-share; half the cost, half the produce. Jennifer and Kyle also offered work shares; for $180 (half-share) or $270 (full), participants could work to earn the rest, by helping in the gardens or during haying.

All of this—and it was their most ambitious program yet—was contingent on getting reliable help in the form of farm apprentices, who were unpredictable. Some of them worked as hard as you did; some of them didn't. Some of them had grown up on farms and knew what to expect; others were from the suburbs or cities, with quite idealized visions of farm life and farm work, and they discovered that they didn't like or couldn't do the physical labor, didn't like to see the animals killed, didn't want to get up that early, didn't like to work that late, were surprised to learn that there really was no such thing as a day off (although Jennifer and Kyle granted one day a week for respite), and were nonplussed by the nominal pay—ranging from $75 to $130 a week, if you were experienced at farm work.

One constant of farming is its inconstancy, its mutability. Without a predictable routine, a farm cannot function, but within that routine are a myriad of elements that do not always behave as expected. The cycle of farming—this particular kind of small farming—imposes its own seasonal and animal rhythms and has its own logic.

Its rewards are tangible, and daily: the food grown, the meat raised, the land the farmer sustains and that sustains them. But in the end circumstances beyond the farmer's control—weather, the economy, a stock market freefall such as after the attacks of September 11—has as much to do with success or failure as anything else. Even the smartest farmer, the one who always seems a step ahead, will admit that there is an inherent mysteriousness as to why something works one year, and not the next.

Jennifer and Kyle's planning for the year to come showed both the virtue and the occasional haphazardness of their thinking. They liked flexibility, they wanted the option to try new things, they had ambition, and they were imaginative. They also tended to leap first, and look later, which Kyle acknowledged freely at the end of 2003, when they sat down to review their management, what had worked and what had not. "I'd like not to be so scattered, not to do everything," he was to say. "We don't always consider the labor involved, which is a cardinal sin," said Jennifer. They both admitted that there were things that had not gone as planned, that had taken too much labor, that had not returned the investment, and thus, they would change or drop them altogether.

"I'm always trying to articulate the reason why we farm," Jennifer would muse at the end of the year. "Obviously we want to stay afloat, we don't want to lose money. If we were really serious about making money, we'd get equipment. Sometimes you have to spend money to make money."

Kyle interrupted. "I have a problem with that, with that cliché. If you want to make money, don't farm."

"My goals for 2003 were to get the house primed and the roof painted, to put money back into this place," continued Jennifer.

"I'm not sure what my goals are, or were," said Kyle. "Pretty much raising Brad. We're probably not a good example of a business plan."

"They're good goals," Jennifer asserted.

"They're nice goals, but if you get to the end of the year and you don't know if you've met them. . . ." Kyle shrugged. "I don't think you want to be logical about farming, or you wouldn't do it. I always felt I'd rather farm than be on Wall Street."

While Jennifer was the one with the passion for animals, Kyle was neutral on the question, tolerant but not dewy-eyed. He had an affinity for the sheep; he admired the way they functioned efficiently as a flock, always moving as one, and he liked the money they reliably brought in. If a lamb was stillborn or died shortly after birth, his face wore an expression that was not quite grief, but regret and disquiet. "It would be nice to know what happened," he'd say, gently fingering the dead lamb's fleece. (When animals or birds died, they were put, as was customary, into the trench of manure and hay that the gutter cleaner would later evacuate, taking it up a conveyor chain to the outside of the barn, where it was then disgorged into the manure pit. It generally took three months for a lamb carcass to decompose.)

But he wasn't crazy about the pigs (capricious, difficult), wished they'd deacquisition their cows (the getting-loose thing), didn't much care for horses (just didn't have much of a feel for them), and hated goats, a prejudice born of a short-lived experiment when they had tried raising them and had ended up chasing them all over the place because they broke free and ended up on neighbors' lands, and because they were needy, continually wanted and expected human attention.

It wasn't that he disliked domestic animals, but that he preferred wild ones—birds, to be exact—and had spent a great deal of time from adolescence on studying them and working to preserve their habitats, as when he had sat near his parents' barn, intently watching the horned larks nibbling on cracked corn he'd put out for them.

Of course, a lot of Kyle's deflating talk—the Yups, Nopes, and Maybes; the absolute refusal, born of observation and experience, to sentimentalize or romanticize anything having to do with farming or nature—was a deliberate counterbalance to Jennifer's ardent proclamations and averrals and general excitability. If she went to the brink in her speech, he pulled her back. Kyle was careful and precise in his wording and thought hard about questions and their answers. There was a clarity in his thinking. He had, if he desired it, the makings of a teacher, in that his explanations of how and why things worked, or didn't, were framed in such a way that anyone—even someone who didn't know anything about farming—could clearly understand without feeling they were being patronized. He was capable; one only had to see him working steadily and quietly and with precision to recognize that.

Jennifer and Kyle were, in their own way, a kind of oddly functioning comedy team; he was George Burns to Jennifer's Gracie, the skeptical Costello to her more frenetic Abbott, Crosby to her Hope. She delivered the setup; he delivered the punch line. She circled around a point; he came straight to it. She was gregarious; he had a solitary, reflective streak. She was exuberant; he was self-contained. She made a fuss of the animals, smooching and cooing, giving them pet names; he didn't. His outward affability and dry wit concealed a certain shyness; she concealed nothing. He was taciturn; she talked constantly. That had been one of the things he'd first noticed about her. "I was captivated by all that chatter," he says.

That, and the long blonde hair and the big hazel eyes and the open, friendly, ingenuous expression and the slim, slight build and the fact that, during presentations at graduate school, she had a certain intensity, a way of furrowing her brow and asking questions frequently and intelligently.

Brad plays under the watchful eye of his mother while his father gets change for a customer at the market.

Now, after seven years of marriage, he had a way of puncturing her pronouncements with a pointed turn of phrase or an eye-roll, a raised eyebrow or a hard stare. Sometimes, just one word was enough, said with the kind of clipped emphasis to let you know that she had, in his opinion, gone too far: "Jenn-i-fer." Sometimes she just pushed him too hard, was too critical, or brought up some complaint from years ago, as if it had happened only the day before and was still open to debate, to which his response might be a joke, or a retreat, physically or emotionally, into silence. The kind of personal revelations that Jennifer dispensed quite casually did not sit easily with him. When she began pulling out old love letters or remembering old boyfriends or talking about something that he considered too intimate for public consumption, the look on his face was one of barely concealed tension and worry and unhappiness.

Was nothing sacred? Just how far would she go, anyway? Why did she keep bringing up the old boyfriends? How much would she say?

Quite a lot, actually. That was her way. When it all got too much for him—everything that was required of him as a husband and father and farmer—he took refuge in his wood shop in the barn, where he turned out wooden bowls and platters from maple and cherry and birch and butternut, some of them from trees that he'd cut down himself on the property. These bowls had started out as the most unprepossessing dusty logs. Only a dreamer, an artist—although he would demur at the description, even mock it—could look at one of these logs and see in its rough bark and blocky, dull heft, the sheen, the grace, the form and line of the thing that it would become.

He was, to those who knew him superficially, most unreservedly himself with his son, Brad: hoisting him easily upon his shoulders, reading to him, pulling him on a sled through deep snow outside the house, pretending he didn't know where Brad was, loudly calling his name, while Brad, hiding behind a chair or in a pup tent, giggled uncontrollably, squealing "Daddun!" when his father grabbed him up in his big hands, and pulled him close in a deep embrace.

His son had opened in him, as happens with many young fathers and mothers, deep and vulnerable and unexpected reserves of feeling that were quite different from the love one feels for a parent or a lover or partner. Brad commanded in him the deepest loyalty and regard and affection, a feeling that was returned in equal measure by Brad. Kyle could discipline when he had to—Brad had once left a plastic syrup container on top of the woodstove, which could easily have resulted in a fire, and that had resulted in a very stern talking-to, one that left Brad shaken by the unmistakable seriousness with which Kyle delivered it—but he was largely indulgent. Not overindulgent, but tender, bemused, and paternal in the best sense of the word. Jennifer was the disciplinarian, the one more likely to lose patience with Brad.

Kyle was six feet tall, imposing, a big man, with hazel eyes, hair that had once been redder and blonder, but was now graying. His family background, way back, was German and Welsh and Scottish, with a dash of French Huguenot, and he had that look that Americans do, of being both unmistakably American and somewhere in their background, unmistakably something else. You could look at Kyle and not think he was anything but American, but if you looked closer—at the ruddy, red-haired, long-jawed, raw-boned face—you could also see a touch of something Celtic, or possibly Gaelic, or if you went back further, Viking.

His mother's side of the family, the Defenbaughs, originally spelled Tefenbaugh, were German Protestants who had emigrated around 1710 from the German Palatine—an area roughly west of the Rhine and north of the French province of Alsace. The migrants from the Palatine, fleeing political instability and religious persecution by a Catholic majority, had made their way to England, where, as Protestants, they had been offered asylum by the reigning Queen Anne, and then to America.

The Defenbaughs eventually migrated to Schoharie, west of Albany, New York, and east of Oneonta, along what is now U.S. Interstate 88. The Defenbaughs—"My grandfather used to say that Democrats spelled it with one 'f' and Republicans with two," observes Lois Jones—were part of a group of some 700 settlers who had been encouraged to settle there by the English, to help harvest tar and turpentine from pine trees. Eventually they drifted into other professions: farming, milling, organ and coffin making.

As part of the migration west the Defenbaughs moved, first to Pennsylvania and then, in a Conestoga wagon, over the Alleghenies and into south-central Ohio. The Defenbaugh farm was, when Kyle's grandfather, William, owned it, sizable—2,800 acres. In 1961, on the death of William—who was the last of four generations of undertakers, a family profession that had had its origins in the coffin making of the eighteenth and nineteenth centuries—that 2,800 acres was divided in two. Half went to his brother and half was split among the four daughters, which amounted to about 350 acres per daughter, with some of the daughters getting more or less acreage, depending on how good the land was deemed to be.

In Vermont, 200 to 300 acres would be considered sizable, something to brag about. In Ohio, 200 to 300 acres was merely respectable; anything less than 200 acres, not even worthy of mention. Sandy and Lois Jones not only worked the land they'd inherited, but took care of some of the land belonging to Lois's sisters, some 600 acres in all.

When the farm went bankrupt in the 1980s, Sandy and Lois Jones deeded it over to their four children, on the condition that they could continue to live there until they died; they live there still, although they are retired. The farm that was, not so long ago, nothing extraordinary or unusual, just one of many such farms, is now something of an anomaly, with subdivisions coming in around them—one- or two-acre lots bought by people moving away from Columbus, from Cleveland, looking for the good life, the rural life.

Sandy had been a forester for International Paper in Georgia, where Kyle was born in 1959. The Jones side of the family was Welsh, and Scottish, with a trace of French Huguenot; Sandy Jones had an ancestor, Duncan McArthur, who was one of the original surveyors of the Ohio River, a representative to the United States Congress from 1823 until 1825, and the eleventh governor of Ohio, from 1830 until 1832. There were relatives who had come into Massachusetts in the 1600s, and then hopscotched west via New York State and into Ohio, settling in the same south-central region as the Defenbaughs. The peripatetic American story.

Another American story: the Megyesis, who emigrated from a village called Csehimindszent in western Hungary to the United States after the end of the First World War and the subsequent dissolution of the Austro-Hungarian Empire. In Hungarian, Megyesi translates roughly as "those who tend to the cherries." In Hungary, the Megyesis had lived on the estate of a rich man, in an arrangement that was close to feudal. He let them live on an acre or two of land, and in return they worked his land and were permitted to keep some of the crops. There seemed to be, says Jennifer's father, Louis Megyesi, no way up or out.

But there was a young woman, Mary Megyesi, who possessed the kind of pluck and independence of spirit that would lead her to emigrate, alone, at the tender age of seventeen, to the United States, settling in Albany, Louisiana, in the Delta, not far from Baton Rouge. There were cousins living there, in a Hungarian enclave called Arpadon. Mary got a job in a store, eventually married the owner of the store, and saved up enough money to send for her parents and her eleven siblings, one of whom was named Geza—Louis's father and Jennifer's grandfather—and who arrived in the Unites States in 1921 at the age of twelve. Geza Megyesi eventually learned a trade, blacksmithing, that took him during the 1930s and 1940s from Louisiana to Ohio, back to Louisiana, from Louisiana to California, and then finally to Michigan, where he became a blacksmith and inventor at the General Motors plants in Warren, and where he lived until the end of his life with his wife, Julia, a Hungarian he had married in Louisiana.

Louis Megyesi met his wife, Beverly Stine, when they were both working at the University of Michigan Hospital in Ann Arbor: he as an orderly, she as a nurse. Beverly's family was of Welsh and English extraction, and had emigrated to western Pennsylvania, in the Alleghenies, ninety miles from Erie. Her father was a railroad man on what was then the Erie Railroad, and when she was young they moved from railroad town to railroad town in Pennsylvania and Ohio. Louis Megyesi and Beverly Stine married in 1960; she was twenty-three and he was twenty-five. They, too, wandered—

Jennifer feeds the ducklings. She has been passionate about animals since she was a young girl.

from Michigan to California to Vermont, a place neither had ever been but had seen in a picture on a calendar at a party in San Francisco, where they were living in 1962. There was a photograph that showed a town in the middle distance. You could see the graceful, white steeple of a church, probably a Congregational Church, the kind of church that said, without even having to think about it, New England. It was nothing if not idyllic looking. The town, when they looked closer, was Worcester, Vermont.

"That's where we should go next," Louis said to Beverly in a fit of inspiration. So they did. Circumstance brought them eventually to Middlebury, where Jennifer was born in 1963, her sister Anna in 1965, and her youngest sister, Laura, in 1969. Louis, who had majored in English at the University of Michigan, and won prizes for his writing, eventually secured a job teaching English at Middlebury Union High School. By 1964, they had saved enough money to buy a house in East Middlebury, where all three daughters grew up.

From a very early age, Jennifer and her sisters were encouraged to care for animals on the theory that doing so would teach them something about life and death and birth, and foster in them a sense of responsibility toward other living things. They had chicks, ducklings, cats, and dogs. They put up a sign outside the house advertising their services as wildlife rehabilitators; charge, one cent. Neighbors took them at their word, depositing with them injured or orphaned blue jays, starlings, robins, cedar waxwings. They cared for two orphaned baby raccoons which, left outside during the day, would at night crawl up onto the roof and down to the window of Jennifer's room, where she, upon hearing a tapping on the window, would let them in and into her bed. Raccoons in the bed, ducklings in the tub with the girls at bath time, a horse, a pony, various stray birds; this was not a household that turned a cold shoulder to animals.

Louis Megyesi had made a place for his daughters near a little pool in the woods behind the house—it was all woods then. He had cleared away the pine needles and rearranged the rocks, and called it Land's End. It was a refuge for the girls, a hideaway. They would disappear there for long stretches of time, the place being safe enough that their parents rarely worried about them. Once, they were gone for a longer period of time than usual, and Louis Megyesi went to check on them. Jennifer and Anna were sitting up in a hemlock tree, holding a long, dangling halter. On the ground beneath them, they had scattered corn, with which they hoped to lure deer. They had sat there for hours, patiently and expectantly, thinking that a deer would surely materialize and when it did, they would slip the halter over its neck.

Jennifer went to Middlebury Union High School, where her father taught English, a situation that had its difficulties, particularly when she got to an age where she showed an interest in boys, and they in her, which resulted, to her acute embarrassment, in her father making inquiries of the other teachers as to the suitability of such and such a suitor. Of herself in high school, she says, "I was a skinny, short-haired, introverted little shit." At the close of her junior year in high school, she went to Brazil as an American Field Service student for one year, from where she returned, in 1982, she says, with "boobs and blonde hair." Photographs bear out her assessment. A passport photograph, taken when she was seventeen, makes her look prepubescent, almost childlike. A photograph taken afterwards shows not a child but a young woman, more assured and less shy. "I had had a year of freedom, and when I came back I was different," she recalls. Plans she had had to be a veterinary technician, which had seemed firm before Brazil, now seemed tenuous. She tried studying art at a state college in Vermont, and then went to the State University of New York at Delhi, in the Catskills, from which she graduated in 1984, after a two-year course of study, with an Associate's in Applied Science (AAS) in Veterinary Technology.

In 1986 she enrolled at the University of Vermont (UVM) in Burlington, to study biology and conservation, and in 1988, as a result of her earlier course work at SUNY-Delhi, graduated with a BS degree. While she was at UVM, she caught sight of a notice on a bulletin board: "Volunteer seabird biologist wanted for remote island in Hawaii. Room, board, and airfare paid for." The notice asked for a ten-week commitment to the project run by the U.S. Fish and Wildlife Service French Frigate Shoals; Jennifer stayed nearly four years, collecting data and writing monographs on a variety of Pacific seabirds, including the brown noddy and the great frigate bird, an unusual creature that

Kyle takes hay out-
side to the horses
on a frigid morning.
He and Jennifer
determine how
many animals they
can keep through
the winter.by how
much hay they have
on hand.

had feet like a chicken and never landed on water. Upon her return to Vermont in 1992, at the age of 29, she applied for, and gained entrance into, a graduate program at the University of Massachusetts, Amherst.

Kyle left the family farm at eighteen, and never really looked back. "I was done," he says flatly. Some farm kids get as far away from the country as fast as they can, seeking out the cities or suburbs, visiting the farm once or twice a year at holidays, and then fleeing gratefully back to civilization, wondering how they—and their families—could ever have endured that life, that constraint, that numbing routine, that insularity. But cities never interested Kyle. Far from it. "I definitely left the farm, but I had no desire for urban life," he says. "It just seems to me like a real unhealthy, unhappy life, and I realize that other people feel the opposite. Good, they can stay there. A city is a nice place to visit every six or seven years."

If one were to analyze his résumé, beyond a recitation of fact, it would show a doggedness, a perseverance, a coming to fruition of his curiosity in the natural world—an inquisitiveness and consideration of how natural systems work that had begun when he was a boy on the farm. He'd worked his way into the national park system, starting in 1979, when he'd made his first trip east to Cape Cod National Seashore, where he worked as a volunteer doing tern counts and putting up signs to keep people away from nesting colonies. In 1980, he had entered the College of the Atlantic, in Bar Harbor, Maine, studied public and environmental policy, and had graduated in 1982 with a degree in human ecology.

From college on, he began to focus on birds and conservation, doing seabird surveys up and down the East Coast, from Maine to Cape Hatteras, for the Manomet Bird Observatory, now the Manomet Center for Conservation Sciences in Massachusetts. In 1983 he was hired as a seasonal environmental-protection assistant at Acadia National Park, in Maine, monitoring air and water quality, and in 1984 was offered a permanent position. He stayed with this until 1988, when he went into a year-and-a-half training program at Cape Cod National Seashore in Massachusetts. Upon completion of the program, he became Natural Resources Manager for Cape Cod National Seashore, a job that entailed, as it had at Acadia, air- and water-quality monitoring, as well as assessment and protection of nesting colonies of the piping plover and the least tern. While there, he helped to organize one of the first negotiated rule-making exercises in the National Park Service's history; an arbitration between people who sought greater access to the beaches for off-road vehicles, and the park service, which sought protection for the piping plovers, whose eggs and young were easily destroyed when off-road vehicles unintentionally ran over them.

While at Cape Cod, he entered into the graduate program at the University of Massachusetts, Amherst, commuting back and forth. When Jennifer first met him, in the fall of 1992, he already had a reputation as a birder of some distinction, with numerous scientific monographs to his credit. "Kyle Jones? *The* Kyle Jones?" she had said to him.

At the time, they were both involved, or had recently been involved, with other people. Jennifer says that from the moment she saw him, she was after him. "He had no clue," she says. He was tall, he was handsome, and, significantly, he was part owner, along with his brothers and sister, of a farm. Her old boyfriends not only did not have farms, they had no interest in farms; or, at least, not the kind of intense interest that Jennifer had, and which she intended to pursue, preferably with someone by her side. "I never wanted to live in a city—I always wanted land. Even if we stop farming someday, I would still want land," she says.

Her sister, Anna, had told one of the boyfriends, Look, Jennifer wants one thing out of life, and that is to go home, to Vermont. If you live in Vermont, you are going to be one of three things: a teacher, a nurse, or a clerk in a store. Or, she might have added, a farmer.

Jennifer and Kyle's courtship was one of fits and starts, and miscues, and enigmatic little notes left in mailboxes, and Kyle appearing, from Jennifer's perspective, to take an interest in everyone but her. There were invitations extended and muddled and finally accepted, and a growing closeness, a mutual realization that this thing between them was significant. By December 1993 they

Kyle uses a chain saw to cut a fallen butternut tree at the Marsh-Billings-Rockefeller National Historical Park. He has been working at the park since 1997, and currently works there part-time as an ecologist.

were discussing marriage, with an eye to setting a wedding date in 1995. "It wasn't bells and whistles," says Jennifer. "I can't explain what it was." This is more than Kyle will volunteer; the subject makes him retreat into a prudent silence.

While they were engaged, Kyle continued working on Cape Cod. Jennifer moved to the coast of Maine, to do bird restoration work for the U.S. Fish and Wildlife Service on Petit Manan Island in the Gulf of Maine, as way down east as you can go before entering Canadian waters and the province of New Brunswick. She worked at culling the great black-backed gull population, an invasive species from Labrador that had been drawn to the Maine coast by feeding opportunities arising from increased development. The gulls harassed adult seabirds, ate their chicks, and usurped territories, and as a result, the native populations of Atlantic puffins, roseate terns, common terns, black ducks, and black guillemots had declined drastically. Culling the great black-backed gulls, the argument went, would help to restore the native seabirds.

Culling is, of course, a euphemism for "killing": the great black-backed gulls were fed a powdered, toxic substance mixed with butter, which caused kidney failure and then death. "The public hates it," says Jennifer. But wildlife management, like farming, requires hard thinking, an array of choices that may range from best to okay to not-so-good, to "Is this something we can all live with?" There are decisions that—while they may be perceived by the general public as agonizing or insupportable or inexplicable (why not just leave the gulls alone?) —must be weighed in the balance, not decided solely on popular emotion. And the balance was, if you can cut back on the black-backed gulls, you have a chance to restore healthy populations of puffins and terns, which, in turn, might lead to the restoration of other species; a hypothesis that was borne out by the success of the project.

After their marriage in October 1995 in Maine, Jennifer transferred from Petit Manan to Monomoy National Wildlife Refuge on Cape Cod. She and Kyle began to discuss seriously the idea of farming. After initially considering Maine, but rejecting it (the price of land along the coast was high, and they felt that growing conditions were not optimal), they focused on Vermont. They began to scour Vermont newspapers, and Kyle zeroed in on the Marsh-Billings-Rockefeller National

Historical Park, which, at the time, was not yet open to the public and did not have an open position for which he might apply. "I started pestering them," he recalled. His persistence paid off; he was hired in the summer of 1997, and moved up right away, living in a motel in Quechee, Vermont. It was shortly thereafter that they saw an ad for a trailer for rent in East Bethel. Jennifer called the owner, a man named Buck Kendall. "We want to move in next week," she told him. "That's fine," he said, not turning a hair. "The thing is," she went on, "we have animals." That was fine, too, he answered: He had this old shed Jennifer and Kyle could use to house them. The following week, Jennifer drove up from Cape Cod in a borrowed Jeep Wagoneer, pulling a trailer behind her in which were one horse, Michael, and one goat, Flag. In the back of the Jeep Wagoneer were forty chickens, six sheep, four cats, two ducks, and one goose. By hook and by crook, Jennifer had come home to Vermont.

When Brad was born on January 4, 2000, Jennifer held him for ten minutes without knowing what his sex was. Dazed and triumphant, exhausted and elated, her only thought was to cling instinctively to the child to whom she had just given birth. "Don't you want to know what it is?" the midwife asked her. "So I lifted his leg up like a lamb and I saw his testicles," says Jennifer. They named him Bradford in honor of a close friend, Brad Kausen, who had died of melanoma not long after he and his wife, Donna, had attended Jennifer and Kyle's wedding. Brad and Donna had lived in Addison, Maine, on the coast, and had embraced the kind of self-sufficient existence about which Helen and Scott Nearing wrote in *The Good Life* series.

The Kausens grew and raised everything themselves; there was almost nothing that they ate or wore that they hadn't produced. Jennifer had lived with them, from the summer of 1993 to 1995, while she was still a refuge biologist at Petit Manan. She had been moved by what she saw as the shared purpose of their life together, and their generosity, and the simple fact that they practiced what she dreamed of doing herself: they farmed in a practical, self-contained manner, without making a fuss about it. They taught her, she says, nearly everything: how to raise and slaughter poultry and livestock, how to can, how to save seeds, how to hunt and tan deer—even how to safely prepare and eat roadkill. They showed her that small farming could be done.

Brad Kausen had died young, at the age of forty-two. It was natural, Jennifer and Kyle felt, that they should name their son for him.

The early months of Bradford Jones's life were difficult. He was jaundiced and had trouble nursing. But these obstacles seemed as nothing when, in March 2000, Jennifer noticed that Brad's hands and feet had developed what looked like a rash, raised spots dispersed across the skin. This might not have been cause for alarm in and of itself, but a few weeks later she noticed that the spots were now on the cornea of his left eye. As the weeks passed, the spots began to spread in the left eye, from the cornea to the iris. His eye began to dull and to take on a gray and lifeless opacity, as if it was the eye of a dead fish.

All Jennifer and Kyle could see of his eye were hundreds of these pinprick white spots, a sort of pointillist nightmare. None of the various doctors consulted over the course of the coming year could tell them what it was, although they advanced diagnoses that staggered Jennifer and Kyle with both their seriousness and their apparent improbability: congenital syphilis, herpes, tuberculosis, leukemia, sarcoidosis, even leprosy. By May 2000 he was, in effect, blind; the brain had stopped sending signals to the optic nerve.

After consultation with a pediatrician and a pediatric ophthalmologist, Brad was put on antiviral drugs and a high-powered steroid, prednisolone acetate, to reduce the inflammation in his eyes and to help restore vision. Despite the steroid, the disease also spread to, and compromised, the vision in Brad's right eye, although he never lost complete vision in it as he had in the left.

As a result of the loss of muscle, his left eye, which had been the first one to show the peculiar spots, crossed into his nose; to keep the left eye working, doctors patched his right eye for eight months. He would eventually require surgery to correct his vision in both eyes.

The only reason he did not lose sight altogether was the application of the steroid. There was still no definitive diagnosis, and in fact, a correct diagnosis would not come until nearly a year after the disease first presented itself. In this time they had seen as many as seven doctors, some of whom were frank in admitting they had no idea what it was, and some of whom were specialists that, Jennifer and Kyle felt, would rather offer any diagnosis, no matter how traumatic, than no diagnosis at all.

The round-robin of referrals from one doctor to another had an unintentional irony to it: the major medical center that was closest to them happened to be the last place they went. When they finally brought Brad to a pediatric ophthalmologist at Dartmouth-Hitchcock Medical Center, the opthalmologist looked at him and said, without missing a beat, "Oh, JXG," or juvenile xanthogranuloma, a rare dermatological disease, its etiology unknown, that occurred mostly among young children and resulted in an acute loss of vision. If it went undiagnosed and untreated, permanent blindness could be the result.

Until the progress of the disease was checked and it abated—it was a self-limiting disease with a life span of three to five years—Brad would require continued medical supervision, treatment, and surgery. All of this was unthinkable without the safety net of health insurance. "If we hadn't had Kyle's health insurance, we would have been sunk," Jennifer says. There had been talk about one of them working full-time on the farm, while the other worked full-time at a job, but this was no longer practical. Giving up both Kyle's federal salary and Jennifer's steady wage from the animal hospital in Bethel was out of the question. (Kyle would make the move to part-time work at Marsh-Billings-Rockefeller National Historical Park in the fall of 2001, so that he could spend more time on the farm.)

When the winter of 2003 began, Brad had just turned three. His sight was fully restored, although he would still need more surgery on the right eye, which drifted upward. Brad had the same blond hair of his mother—in summer, it looks as if it has been burned white by the sun—and, when he was feeling stubborn or pugnacious, the set of his father's jaw. It was possible to see his grandfather, Louis Megyesi, in the broad planes of his face. His voice, the inflections and elocution, was uncannily like that of his mother, a higher pitch that rose on the last syllables of words. He had some of the reserve and also the directness of his father. His round face and bright, deep brown eyes and the straight line of bangs across his forehead gave him the look of a little owl. His frame was slender and wiry, like his mother's. He had a certain verbal precocity, a sense of playfulness:

"I love Mom the most. I love Daddun the least," he would say.

"Whaaaat!!!" Kyle would exclaim in mock outrage.

Brad would then immediately reverse himself. "I love Daddun the most. I love Mom the least."

He had just become aware that when animals were taken away from the farm, they did not come back, and he did not like it. "Those are my pets," he protested, sometimes howling and in tears. It was explained to him that because they raised the meat they ate, they had to kill the animals or birds. He was endowed with a furious energy that kept him running and playing, darting like a dragonfly from barn to house to fields. When Jennifer or Kyle went down into the milk house or the fields, he went with them. He accompanied them into the barn, although he was not fond of the stink, and was not shy in telling his mother or father that they smelled when they came back in from doing chores. When they collected maple sap at sugaring time, he was often there beside them in the truck or on the tractor.

When Jennifer went to the greenhouse in South Randolph, he came with her and played on the dirt of the greenhouse floor. He liked to monkey around with the hose, which brought a stern warning from his mother, or clamber up onto the tables and jump from one to another, which brought another warning. In early spring, he'd fling himself into the puddles on the driveway, and come up looking as sodden as a soccer player on a muddy pitch. He spent at least one day a week with Louis and Beverly Megyesi, who took care of him either at the farm or at their house in East Middlebury. He was their first grandchild, and they doted on him; he doted on them in return, and called them Mima and Bipa.

After careful deliberation, Jennifer and Kyle had decided that Brad would be their only child. The responsibilities of the farm were such that they could not see caring for another one.

In early January, Jennifer attended a meeting in Royalton held by NOFA, the purpose of which was to discuss the USDA's recently implemented organic standards. Would Vermont's organic farmers accept them or continue to maintain the organic standards as outlined by NOFA and certified by Vermont Organic Farmers (VOF)? The NOFA standards, many of the farmers felt, were more scrupulous and just that much closer to home.

There was considerable suspicion of the USDA. There were concerns that the big organic operations in states like California, organic labels notwithstanding, were not so far removed from corporate farming, that somehow the USDA would weaken, not uphold, standards in favor of big business. That the small farmer, as usual, would be left holding the short end of the stick.

(This perception, which sometimes approached the pitch of conspiracy theory, was not without basis. A month later, a public outcry would arise over the discovery that exemptions that would have considerably diluted the national organic standard—rendering it organic in name only—had been slipped into the $397 billion Omnibus Appropriations Act. The furor was such that, after an initial period of silence, Ann Veneman, Secretary of Agriculture appointed by President George W. Bush, issued a statement in which she averred that, to ensure "the integrity of the organic label placed on consumer products," it was critical to uphold the standards that had been implemented in October 2002.)

There was also grumbling at the meeting that the USDA was the 800-pound gorilla, throwing its weight around simply because it could—or the bull in the china shop, blundering in where it had no business blundering, and with little feeling for all the groundwork that had already been established independently of them. The general sentiment seemed to be that if Vermont had a political and business climate congenial to organic farming, it was no thanks to the USDA.

"I'm a Vermonter and I'm proud of it," one old-timer said defiantly. "The USDA has done a lot of screwing up. Vermont hasn't done a lot of screwing up. Let's keep the VOF label."

Such defiance isn't unusual; farmers have a culture of defiance married to an innate populist streak, regardless of their politics. They feel their backs are always up against the wall and that the

Brad enjoys a roll in the cold water and mud. Kyle later brought him into the house and gave a fully clothed Brad a shower.

forces arrayed against them, whether natural or man-made, are too powerful, and so they complain loudly and often bitterly.

This is particularly true of small dairy farmers, who catch hell coming and going, selling at wholesale prices and buying at retail—the reverse of how business tells you it should go. But to an extent, all farmers are united by their distrust of the government, the banks, the middlemen, all of whom are judged to be—if not all the time, then some of the time—meddling and obstructionist and rigid, with all the imagination and empathy of a committee of Soviet apparatchiks dictating policy.

There were some twenty people arranged around a horseshoe-shaped table in the main room at Royalton Academy, a former school turned community building. It was a respectable but not huge turnout—but then again, it was 5 degrees above zero outside. There were back-to-the-land farmers who looked as if they were right out of Central Casting—peace symbols and NO WAR IN IRAQ buttons and bottled water. There were farmers, hulking and shambling and utterly silent, who looked like dairy farmers desperate to find a way to keep their farms going, and who'd heard that you got more money in the long run if you went organic. There were farmers who looked like the preppies they had once been—moccasins and corduroys and plaid shirts. And there were farmers who had come just to listen.

That there were people here at all on such a cold day said something about the strong feelings people had on the issue, given American farmers' history of shying away from large-scale organization. The Grange movement and the kinds of cooperatives that NOFA represents are, after all, voluntary. There are unions for steel workers and auto workers and garment workers and migrant workers and hospital workers and truckers; there are no such national unions for farmers, who are physically scattered across thousands of miles of land and have tended toward political conservativism, initially suspicious of organizations that, as they saw it, smacked of socialist or communist or anarchic sympathies.

Of course, given the history of child labor on farms, farmers may not have wanted to organize because they would then be subject to the same level of safety and health standards that attached to other industries. But one man's child labor is another man's family business, and no business of yours. Farmers are, by and large, a perversely independent lot, dedicated to pursuing happiness in their own individual way, in a country that asserts—in theory, if not always in practice—the rights of the individual above all else.

John Cleary, a slight, dark-haired, bearded man who is the VOF Administrator & Technical Assistant Coordinator, had spoken of the need for farmers to agree on what the guidelines should be. Republican Jim Douglas had just been sworn in as the governor of Vermont, succeeding the Democrat, Howard Dean. Douglas knew farming, was sympathetic to farmers. Here was an opportunity that should be seized.

"We have a new government in Vermont, a new Committee of Agriculture, and it's really important for farmers to voice opinions immediately," he said, adding that, unfortunately, the "Executive Committee of VOF couldn't have certified farmers on the Executive Committee, due to a conflict of interest stipulation."

"Isn't that stupid?" Jennifer blurted out. "The USDA says you can't be an organic farmer and be on the committee."

Cleary had run down the numbers, and they were encouraging. "There's continued growth in both certified and non-certified organic farming. There's been incredible growth in the number of farms showing interest in 'transitioning' to organic." There were, he said, currently 23,202 acres in certified organic production. The previous year had seen $15.5 million in sales of certified organic goods, and $12.2 million in income for certified organic processors, a total of nearly $28 million. People were, he added, "waking up to the fact that this is the fastest-growing area of agriculture in Vermont."

"We've had to adopt the National Standards," he continued. "We're not allowed to have additional standards, but we're allowed to have guidelines as long as they're not in conflict with the national standards. We still have a role to play in interpreting and implementing standards. There's vague language in the national standards that they've made clear they want us to fill in."

Jennifer pours a fish emulsion mixture onto young seedlings in the farm's rented greenhouse. The plants need watering at least once a day, and are fertilized every three weeks.

The consensus was, said Cleary, that "for important issues, we can't just ask USDA what the answer is, because then they'll give us an answer and we'll be stuck with it. If you say organic producers have reached a consensus, USDA probably won't fight it. They're not interested in conflict."

It was important, stressed Enid Wonnacott, executive director of NOFA-Vermont, that the psychology of the consumer be addressed in whatever steps they decided to take. "Consumers don't know the difference between the USDA organic and NOFA-VT organic label. They have a really big capacity to be confused, and we have to be really clear." To that end, she had written to U.S. Senator Patrick Leahy, one of Vermont's two senators, a Democrat, and a ranking member on the Agriculture Committee. "Consumers have a belief in the VOF label," Wonnacott said.

One farmer expressed what seemed to be the general sentiment of operating within the USDA organic standards guidelines but holding the USDA, philosophically and logistically, at arm's length, of complying just enough that it didn't come snooping around and make trouble; a reasoning that may have attributed more interest on the part of the USDA than it actually had.

"If the dog's asleep," he said, "don't kick it."

One morning, toward the end of January, Jennifer and Kyle sat in their living room waiting for a guy named Russ to show up. Russ Patton was twenty-seven years old and an apprentice at Luna Bleu, another organic farm in South Royalton, a much larger operation that had developed quite a following at the Norwich Farmers' Market. Where Fat Rooster had but one greenhouse on the property, which they would use as a chicken coop come spring, Luna Bleu had five. Where Fat Rooster might hire one or two apprentices, Luna Bleu hired four or five. Where Fat Rooster cultivated two acres for crops, they farmed perhaps three times as much.

Russ had started working at Luna Bleu in the summer of 2002 and had decided to stay on, although, as a southern Californian, he was flummoxed by and unhappy with this winter's cold. He came in looking like a cross between a nineteenth-century woodsman and something out of *Dude,*

Where's My Car?—unshaven, hair sticking out from underneath a baseball cap that read GRUBCO, Buddy Holly glasses fogged over from the cold, red-checked flannel shirt and heavy woolen pants. It wasn't that he didn't know how to dress for the cold, but that it wore him down.

Russ wanted to earn some extra money and Kyle needed help with maple sugaring. Russ had already bartered with them for a piglet and was taking care of that, raising it for meat at Luna Bleu. Kyle and Jennifer liked him enormously. He had a sneaky sense of humor. He said things like "Dude," and "Yo," was, superficially, a slacker, with a slacker's disinterest in convention, and a self-deprecating, sheepish way of moving through the world. Only he didn't slack, he worked hard.

He played with Brad as a child would, with an all-encompassing absorption in the task at hand, as if the outside world, the world of adults, did not exist, and was not quite as compelling as the world of children. How Brad reacted to, and interacted with, people who helped on the farm was, in fact, a critical litmus test as far as his parents were concerned. If Brad liked someone, that person was encouraged to stay or to share meals with them; if he didn't, that person's tenure was likely to be brief. Brad adored Russ, became almost giddy when he came over to visit.

Russ shared with them his catalog of taped episodes of *The Simpsons,* and sometimes came over to watch *Survivor* with them on Thursday nights, as did Kathleen O'Boyle, an apprentice who had worked for Luna Bleu the summer of 2002, and had been persuaded by Jennifer and Kyle to come back to Vermont this year to help them for six weeks during sugaring season.

Jennifer had tried once to get onto *Survivor*—three times, actually, but the last two times she had missed the deadline. An application for the show dangled haphazardly from her cookbook shelf, right next to *The Farmhouse Cookbook, The Chicken Cookbook,* and *The Political Palate: A Feminist Vegetarian Cookbook* by the Bloodroot Collective. No one around her really understood why she would want to submit herself to *Survivor* when she was already living it, although it was obvious that she relished the idea of adventure.

The same held true for her later obsession with applying to be on the public television show *Colonial House,* in which families would attempt to live as settlers had in seventeenth-century New

Brad helps his parents clean organic eggs for sale. At top production, the farm's hens are producing five dozen eggs a day.

England. Here she was, ensconced in a farmhouse in the country and farming in an old-fashioned way; short of using an outhouse or lowering a bucket into a well to draw water, her existence straddled two worlds, the modern and the not-so-modern. She and Kyle did not live as the Amish did. They drove cars and trucks, had the modern conveniences, watched television and scrolled Web sites on the Internet (Kyle sold some of his wooden bowls through his Web site), but in a sense their way of doing things was a throwback, in that it involved manual labor, with only occasional recourse to equipment, and borrowed equipment at that. They did every day what the participants in *Colonial House* or *Survivor* did for a finite period of time. Why did she have to prove it on television?

For *Survivor,* she had sent in an audition videotape of herself in a tankini, midriff bared, with a heavy application of black kohl around her eyes, plucking a dead chicken, the blood from which smeared on her cheek when she put her hand up to push the hair away from her forehead. What could be more *Survivor*-like than chicken blood drawn across one's cheek like war paint? She had street-cred, she thought. She'd spent all that time in the Pacific, on an island, just like the contestants, only she'd done it for years at a time, not weeks. She'd been rugged, eaten rugged, done rugged things. Why wouldn't they want her? But, apparently, they didn't, which she glumly put down to her age and lack of cleavage.

There was a woman on *Survivor* who looked anorexic—you could see her ribs, every part of her looked skinny and wrung out, except for her breasts. What kind of woman, asked Jennifer triumphantly, would be that skinny and not lose weight in her breasts? You could actually see below her skin to the skeletal structure beneath, but then you were confronted with these round, plump, perfect things. They couldn't be real, had to be implants.

She and Russ would curl up in front of the TV and analyze the woman's breasts, and root for a woman named Christy, who happened to be deaf, and make snide comments about another man who, Jennifer said disapprovingly, and tellingly—given the network of neighbors on which she and Kyle were dependent for success on the farm—"made an alliance and then broke it."

Kyle didn't care much about *Survivor* one way or the other, and ignored Jennifer whenever she went off on her *Survivor* tangent, or when Jennifer and Russ discussed, with as much animation as if they actually knew them, the minutiae of which contestant was the more worthy of the *Survivor* title. Kyle had one purpose in mind: he was hoping to enlist Russ to help him with both the job of repairing pipeline, which took place in January and February, and gathering maple sap, which could happen anywhere from late February through early April.

Pipeline was not oil pipeline, but maple sap pipeline. You could drive down almost any back road in Vermont at this time of year and see in stands of sugar maple—what was called a sugarbush—long skeins of blue or green plastic hose winding and crossing from tree to tree in a kind of elaborate cat's cradle, indicating that these trees were going to be tapped. Pipeline was considered a more efficient method than the old-fashioned tin bucket and spout method because it required less maintenance and cleanup, and that work could be dispersed over the course of a year, rather than being crammed into a six- or eight-week sugaring season. There were still plenty of old-timers who clung to their tin buckets and spouts, though. You could see buckets hanging from the weathered sugar maples that lined the dirt roads and which, in summer, provided stately, towering avenues of shade.

Tapping entailed drilling a hole a few inches into the bark of the sugar maple and inserting the tap; each tap funneled the rising sap into the pipeline and from there it ran into larger sap tanks that served as central collection points. (Another advantage of the pipeline and sap tank method was speed: pumping the sap from the central collection tanks took minutes rather than the hours consumed by going from bucket to bucket.) There might be, in the sugarbushes in which Jennifer and Kyle tapped—Charlie and Debbie Morses', Ella Hyde's, George and Agnes Spauldings', as well as other neighbors'—any number of these sap collection tanks, with capacities of 550 gallons, 325 gallons, or smaller 55-gallon drums.

When it came time to empty them, Kyle drove a tractor, onto which was hitched a trailer bearing an enormous tank, into which he pumped the contents of the various sap tanks. He then hauled the trailer and tank down to George and Agnes's sugarhouse and pumped the sap into a separate storage tank on top of the building. When they were ready to boil the sap down, it was funneled

As Kyle pumps out a sap tank across the road, sap drips into a bucket. When sap is running quickly from the 900 taps, friends and family help out. Kyle can make as many as five runs a day with the tractor to collect sap to make maple syrup.

from the storage tank into a 3½-foot-by-12-foot stainless-steel evaporator inside the sugarhouse. Both the gathering of the sap, and the boiling, continued over the course of the entire six- to eight-week sugaring season.

George and Agnes had worked out an arrangement with Jennifer and Kyle. George and Agnes were getting on: he was seventy-four and she was sixty-six, and neither were eager to toil up the steep hills in deep snow and bitter cold to do the laborious chore of repairing pipeline and inserting the taps. If Jennifer and Kyle would undertake the work of repairing pipeline and gathering sap, George and Agnes would gather the 15 to 18 cords of wood needed to fire the evaporator, and they would boil the thin sap down into the highly concentrated syrup—Agnes, who had learned to boil from her father, was quite proprietary and punctilious about this aspect of the operation—and take care of much of the cleanup. Both couples would go in on buying the empty tins that would hold the syrup, and shoulder the cost of equipment, and whatever that year's yield of syrup was, they would share it. The division of labor wasn't rigid: sometimes George and Agnes helped gather sap, sometimes Kyle and Jennifer helped boil, and the assumption was that during sugaring, all would be available to lend a hand and to offer moral support, if need be. It was a fair arrangement, and not the only one on which George and Agnes had embarked with their neighbors: in return for plowing and caring for Ella Hyde's fields, they grew feed corn for their cows on her land, because they had no cropland of their own.

George and Agnes had built the sugarhouse on the hill overlooking their farmhouse in the fall of 1999. It had the look of a dacha in the Russian countryside, with a pointed roof and a stovepipe chimney from which smoke billowed forth during the sugaring season, and an attached shed in which they piled wood. Sugaring was something most farmers relied on to make money in the no-man's-land between winter and spring. How many ramshackle farmhouses had hand-lettered signs out front advertising Vermont maple syrup or fresh eggs, on roads where it was plausible to think that no tourist would ever venture? Too many to count.

You couldn't imagine that sales from maple syrup would be enough to tide a small farmer over in winter, but often it was, or at least it gave the illusion of doing so, if you didn't calculate the labor involved in maintaining the pipeline and gathering and boiling the sap. If the sap was running furiously, as it had in 2002, sugaring could go on for weeks on end, with no time to catch your breath, and everyone was expected to pitch in. "I would cry when I got back from work and see the smoke pouring out of the sugarhouse," Jennifer says.

To anybody not from here, the relentless gray and snow flurries of late February and early March was still winter. But you could read the signs that winter was softening in the more temperate winds and the beginning of melt on the river ice, in the mud on the back roads and the lingering daylight after four-thirty P.M., when just six weeks earlier it had been pitch black by four P.M. It might be two more months before the snow had disappeared for good, receding like a glacier, leaving behind four or five months of accumulated detritus, but it was the beginning of the end.

Before Jennifer and Kyle could get to the sap-boiling stage, however, they had to repair the pipeline, and have it ready by late February; it was possible that the sap might run that early. Usually you thought of March when you thought of sugaring, but George, who had a lifetime of experience in these matters, had drummed into Kyle that you had to be ready to go at any time. Typically, over the course of a year, the pipeline accumulated wear and tear. Squirrels ate through the hose, trees fell on the lines, deer and mice inflicted damage, and in places where the lines crossed well-worn coyote trails, the coyotes, who did not like obstructions in their path, chewed on them.

The challenge lay not in the repair, which was not complicated, but in the number of taps, and in the fact that they were scattered over acres of land. Kyle didn't relish doing all that work by himself, given that there were thousands of feet of pipeline and 900 taps, and Jennifer wasn't keen on it either, because it would mean bringing Brad with them, and there was only so much exposure to the cold and snow he could stand.

"We can't drag Brad up that hill," she said emphatically to Kyle. Russ was sitting at the table with them. "We did when he was just born. Do you remember nursing under the pine trees?" she asked Brad, who was puttering around with his toys. "Hemlocks," Kyle said. "Your mother gets confused." Jennifer looked at him with some irritation.

Jennifer rolls back a muslin and felt strainer while filling a half-gallon container with maple syrup.

Russ was game. "I wouldn't mind being out gathering," he said. "You don't need two people sitting around boiling sap."

Jennifer threw out a number. Twenty dollars a day and a gallon of syrup. Russ tilted his head to one side, his chin dug into his shoulder, and his nose wrinkled.

"I'll have to think about it," he said. "I was thinking half money, half syrup: ten dollars in cash; ten dollars in syrup. More like a barter thing. To be honest, for the amount of work that goes into sugaring, it doesn't seem like a money-making thing."

"Oh, definitely," Jennifer replied. "Agnes says it's what you do in February and March, when there's nothing else to do."

They agreed to do it as Russ would prefer; payment mostly in syrup and a very modest stipend and meals, if he happened to be there at mealtime. "In addition to syrup, food, and money, it's one beer a day, right?" he said with a grin, eyebrow raised, head tilted.

They made more small-farmer small talk: The price of peppers imported from Israel at a big-chain supermarket ("Israel!" Jennifer marveled. "How much did the farmers get for those? It can't be lucrative, can it?") They swapped saved tomato seeds, which led to an exchange on eating tomatoes out of season, which January emphatically is: "I don't really want to eat a tomato right now," Russ said. "Why would I want to?"

They discussed the killing of chickens. "I'm going to learn how to kill chickens this year," Russ said. "I tried one last year but I didn't do it right and it freaked me out."

"Killing the chickens is a hard thing," she agreed. "It's harder for me to do that than the lambs. I have more of an affiliation with the chickens. When you cut chickens' throats, they close their eyes."

"Turkeys are a bear to kill," Russ observed.

"We pith all the birds. It's supposed to scramble their brains. It's psychological. I don't know, I don't like killing birds," Jennifer said.

Then they all piled into Jennifer's car and went down the road to the Spaulding farmhouse. George and Agnes had a dairy herd of some sixty cows, in a brown barn across the road from the farmhouse. George had been born in this farmhouse and lived his entire life there. He had that look that some older farmers do, of having worked so hard their whole lives that they might be anywhere between sixty and eighty. He was angular and bony, and his skin was deeply lined and loose on his thin frame. His eyes were a bright blue, and rather mischievous.

Agnes had cropped white hair and glasses and round cheeks and a girlish giggle, a chortle, although it was also possible to see in her the sterner stuff of the mistress of an estate, who, for all her outward grandmotherly softness, had very specific ideas on how things should be done.

They sat around the dining room table: Kyle next to George who was next to Russ who was next to Agnes who was next to Jennifer: men on one side, women on the other. The ceiling was stamped tin, the floors hardwood, the furniture unassuming. A television in the other room was tuned to the Discovery Channel, to a program called "Hunting the Crocodile." Everyone walked around in stocking feet; boots were left by the door to keep dirt and mud to a minimum. George had a hole in one of his socks. There were bird feeders outside the window; chickadees swooped in and out. Russ stared at them from time to time, when his attention wandered away from the topic of sugaring. Brad had been given toys to play with.

There was, of course, moaning about the miserable January weather. George informed them that his milk pump had frozen overnight. Jennifer said, "Thank God your gutter cleaner hasn't frozen up yet." Agnes brought them around briskly to the matter at hand: how many empty tins, and in what size (pint, quart, half-gallon) were they going to order? It was hard to tell now what the season would be like. If the cold continued as it had been, it would not be warm enough for the sap to build. If, on the other hand, the sap was as plentiful as it was last year . . . ?

There was a pregnant pause.

"We ran out of wood last year, while boiling," said George, passing his hand over his face in a gesture of recollected fatigue. "We cut wood in the rain . . . and the snow . . . and our neighbors brought us wood and . . ."

"The sap kept running," said Jennifer.

"Running and running," George groaned.

"We'd like to buy more pint cans," said Jennifer. "We don't sell that many half gallons or quarts."

"Last year," said Agnes, looking down at her tally, "you sold one hundred and fifty-six pints."

A prolonged and slightly confusing back-and-forth ensued over what was sold, and how much. "This year we got down to almost zero cans," Kyle reminded them.

"I knew I shoulda brought my calculator," Jennifer said.

"There's an adding machine in the other room," Agnes pointed out.

After more adding and subtraction, done by Agnes with pencil and paper, and with Jennifer calculating her own tally, Jennifer announced that they would owe George and Agnes $458.70 for the purchase of enough tins to get them through the season, which also included some equipment needed to upgrade the evaporator.

"I want to get another thermometer," Agnes said firmly. The thermometer measured the readiness of the syrup for taking off. At present they only had one thermometer, which posed difficulties when, to prevent burning or scorching of the pans, they reversed the flow direction of the sap in the evaporator every few days. Problem was, until they could plug the hole in the tank left by the absence of the thermometer, the sap typically poured through onto the floor, creating a sticky mess.

The obvious solution was to have two thermometers, one for each side, rather than to have to undergo this Laurel and Hardy routine of switching sides and straining to insert a plug before all hell broke loose. But for various reasons—thrift, most likely—Agnes had held firm at one thermometer, until it became apparent that it would no longer suffice.

With hand cupped over mouth, eyes glittering with amusement that she had at last given in on this question, George said, in a stage whisper that drew laughter, "Thank God!"

One day in early March, Kyle and Russ climbed up an incline into the sugarbush that belonged to George and Agnes, just off Morse Road. This morning, they followed a ragged trail that Kyle had already made, two feet, sometimes three feet deep in the snow. Sometimes they were able to walk on the snow's crust; other times they broke through. Logs and downed tree limbs lurked underneath the snow, and sometimes they tripped over them. Deer tracks were plentiful; in this winter, the deer had had to hunt long and hard for food buried beneath the snow—you could see where they had pawed through the snow to the ground, in search of vegetation—and many would die. In the spring their bones, flesh and fur still attached, would lie camouflaged against the matted leaves and roots exposed by the melting snow.

This was certainly not the steepest hill that Kyle and Russ had had to climb; in fact, it was relatively easy. The hill behind Chet and Betty's house was much worse: straight up to a ridge from which you had a panoramic view up and down the valley. Russ called it the SOB Hill. That climb would take the wind out of you even without the snow; with the snow, it was a difficult upward slog, particularly when you were carrying equipment and the temperature was hovering at ten degrees.

This hill today was close to the road and it was on a gradual upward incline, easy to get to, and easy to climb. The temperature favored them; it was in the mid-20s. There was something heartening about getting out of the house and into the woods in March, feeling that the worst of winter was behind you and that spring was not far off, which, when you'd lived here long enough, you began to sense in an almost animalistic way, feeling it instinctually before it was obvious to the eye. There were discrete but telling signs that winter was loosening its grip: a migrating flock of cedar waxwings, nearly thirty of them, scattered throughout the branches of a sugar maple, even as it snowed; deer, raccoons, weasels on the move, crossing roads and highways, coming out into the open in search of food; scudding clouds that brought not snow, but rain; Canada geese flying north and west in long, tattered ribbons.

An abundance of sugar maple trees needed to be tapped that morning. Many of these trees had been tapped the previous year or the year before that. You could see the tap holes; they had begun to close over, in the same way a flesh wound heals, leaving behind a pucker of newly grown skin. You did not want to drill a tap hole too near the previous year's tap hole; a maple can only take so

Kyle strings new pipeline from one maple tree to another on a neighbor's land.

many taps. Kyle had both a hand drill and a battery-operated drill, which would have been faster, but—typical luck—the battery-operated drill stopped working. So they trudged back down the hill, and drove down to George and Agnes's farmhouse to borrow a second hand drill. Russ and Kyle worked separately, each to a tree, and often far apart; there was not much in the way of conversation, except shouted questions from time to time, from Russ to Kyle.

They each put their drill up to the bark, and began to turn it by hand; the drill wormed its way into the tree, excreting curlicue wood shavings. Each man drilled 2½ inches into the tree, and then reversed, turning the drill in the opposite direction, rather like a corkscrew, to bring it back out, giving it a last tug to remove it entirely. Then they hammered in each tap. Over the course of three or four hours, Kyle and Russ tapped 180 maples, trudging up one hill and back down again. During tapping, which took place over a period of two to three days as soon as it became clear that the sap was going to run, this was the kind of work they did daily.

This year, it was the third week in March—late, late, late—before the sap began to flow, just around the time that George and Agnes had arranged, as part of a statewide maple syrup promotion, to hold an open house, the weekend of March 22. George and Agnes would, of course, be there, as would Jennifer, Kyle, and Brad, and assorted friends and neighbors. The expectation with an open house of this type was that the sap would have already run in abundance and there would be dozens of pints of syrup for sale. That was not the case this year. This was the first time George and Agnes had held an open house, and the number of visitors was small, perhaps because the weather was uninviting, raw and wet and gray.

Outside, the ice on the Second Branch of the White River was breaking up; you could hear the water rushing downstream in a torrent for the first time in months. In the sugarhouse, the sap was boiling away furiously in the evaporator, steam rising, and the place had the distinctive warm, sweet smell of maple. Jennifer helped Agnes pour off the syrup into the tins; Kyle helped George stoke the boiler with wood. It was an all-day affair. Visitors huddled around the evaporator, which gave off heat, or ate the doughnuts that Agnes had made, or dug into their pockets for cash to buy pints of syrup. Brad liked to drink milk with syrup in it, and he liked the sugaring tradition that called for downing a cup of warm syrup—in his case, a small Dixie cup—and then following it with a sour pickle. Sweet and sour, sour and sweet.

The season was shaping up to be less than stellar, which plunged George and Agnes into gloom. They looked dour. The later the sap runs, the darker it is when it is boiled down; tourists tend to prefer the lighter-colored syrup—what is called "fancy"—to the darker grade, which many people think has the more intense flavor but which is usually relegated to cooking, and not sold as a gift item. Because George and Agnes were not making as much syrup, had not yet begun, really, the open house had the feeling of anticlimax, of expectations deflated.

"Last year at this time there were more than a hundred gallons," said Jennifer. "Now we have fifteen."

"No thaw this January," Agnes said. "Didn't have a thaw until March."

Was the season disappointing so far? George raised an eyebrow. "That's an understatement," he muttered.

Jennifer had arranged with Kyle that he would stay at the sugarhouse with Brad while she would go the greenhouse in South Randolph, to continue the work of transplanting seedlings.

She was eager to leave the sugarhouse behind, and jumped into the pickup truck. On the way there, driving the 7 miles from the farm to the greenhouse, over deeply rutted, boggy roads, she stopped to check a tin bucket at a neighbor's house, to see if there was any sap. There was, a little. She frowned. They wanted more sap, to be sure, but not today. Today was not convenient, but then nothing about sugaring ever was. "I'll tell Kyle, but not George and Agnes, because then they'd want us to get out and collect," she said.

Outside was snow, slush, dirt, raw damp, and unrelieved, dismal gray. Inside the greenhouse was tropical warmth and hundreds of seed trays, laid end to end, table to table, bearing thousands of

Jennifer waters plants in the rented greenhouse. Watering is usually done midday, but she and her family were also doing some planting at the end of the day.

translucent green shoots, which were of varying height, and the only source of color in the place. Jennifer and Kyle, and occasionally Kathleen O'Boyle, took turns going to to the greenhouse daily in February and March, to regulate the temperature and to water and transplant the seedlings. Often Brad went with them. When it was too cold to do anything else, the 70-degree temperatures felt blissful and the fragile spears of green poking up from the dirt in the seed trays gave you hope that winter would end.

Jennifer had brought her two geese—a male and a female that she had had for five years—from the farm to live there while they leased the greenhouse. They ate the weeds and scratched in the dirt and patrolled the perimeter with their waddling gait and snaky necks and their occasional hisses of indignation or alarm. There was nothing to guard against here—once, near the barn, Jennifer had seen them run off a marauding fox—but they were nonetheless vigilant and alert to strangers in their midst.

"Bit the dust, bit the dust, bit the dust," murmured Jennifer, bending over the trays to inspect the seedlings, pulling out those that had withered away to nothing. You could run your hand over the shoots and feel the emergent life; Jennifer called it "tracking the progress of spring." There were the beginnings of leaves, heart-shaped, spear-shaped, star-shaped. In three to four months, every single one of these shoots would have been transplanted into the earth.

There were artichoke, basil, beans, beets, broccoli, cabbage, carrots, catnip, cauliflower, celery, celeriac, chard, chicory, chives, cilantro, corn, cucumbers, and dill. Eggplant, fennel, greens, kale, kohlrabi, leeks, lemongrass, lettuce, melons, onions, oregano, parsley, parsnips, and peas. Peppers, pumpkins, gourds, radishes, rosemary, sage, savory, spinach, squash, and turnips. There were fifty-six different kinds of flowers, among them zinnia, statice, delphinium, snapdragon, salvia, and stock.

The plant varieties had ornate and improbable names like Royal Burgundy (bean), Bull's Blood (beet), Greenwich (cabbage), Bianca Riccia (chicory), Lucullus (chard), Alibi (cucumber), Calliope (eggplant), Zefa fino (fennel), Mizuna Mustard, Red Russian (kale), Gigante (kohlrabi), King Richard (leek), Ambrosia (cantaloupe), Lollo Rossa (looseleaf lettuce), Ermosa (butterhead lettuce),

Deer Tongue (mesclun), Green Arrow (shell peas), Boldog Hungarian (a paprika pepper), Amish Pimiento (pepper), New England Pie and Warty Tan Cheese (pumpkin), d'Avignon (radish), Zephyr and Seneca Supreme (yellow squash).

There would be twenty-eight varieties of heirloom tomato, some of which didn't sound like tomatoes at all: Yellow Pear, Pink Plum, Garden Peach, Hawaiian Pineapple, Persimmon, Cosmonaut Volkov, Brandywine, Black Prince, Green Zebra, White Wonder, and the oddball Mortgage Lifter (so-called because, in a possibly apocryphal tale, the man who created the hybrid was so successful that he was able to pay off his mortgage with the proceeds from the sale of the tomatoes). The tomatoes would not look much like the uniform tomatoes in the supermarket. They would be pink, yellow, orange, red, purplish, and white. Many of them would be awkward and ungainly in shape, not perfectly round, and some would be huge, as much as 2 pounds each, and ugly. Some would be wonderfully sweet and full in taste, others more acidic and tinnier. Some would be particularly susceptible to pests, others less so.

The public was curious about these idiosyncratic varieties, seeing in them a return to an earlier era in which food was not of a uniform taste or appearance, and the farmer was personally known to the buyer. At farmers' markets, the customers peppered Jennifer or Kyle with questions about the relative virtues of, for example, the heirloom tomatoes. Was this one better for paste? That one more suitable for salad? What would Hawaiian Pineapple look like when it grew to maturity? Did it really look like pineapple, or just taste like it? Both Jennifer and Kyle had learned enough about the varieties they grew to talk about them with authority, and enthusiasm. Jennifer felt almost as fondly toward her flowers and vegetables as she did toward her animals, and had little trouble summoning an anecdote or recipe for nearly each and every item on sale at the market.

A customer might wander in, ask some questions, and wander out without buying. The following week, the same customer might wander in again, peruse more closely, and buy one or two items, maybe a tomato seedling, maybe some fresh eggs. The week after that, that customer would make a point of looking at the Fat Rooster booth to see what new produce or goods were for sale. This

One of Kyle's Warty Tan Cheese pumpkins was awarded the blue ribbon at the World's Fair in Tunbridge, Vermont.
Kyle saved the seed for the pumpkin from another he bought at a pumpkin show in Circleville, Ohio.

was how Jennifer and Kyle built the customer base, one interested consumer at a time. If the work in the greenhouse began to seem tedious or overwhelming, they reminded themselves that the frail Amish Pimiento peppers or Royal Burgundy beans over which they were laboring would tender them dividends some six months later, when they lay, in lustrous abundance, on the shelves of the Fat Rooster booth at market.

THE ANIMALS

At the end of February, a man named Dave Hinman came up from Walpole, New Hampshire, to shear the sheep. He liked to say of himself that while he might not be one of the best shearers in the world, he was one of the fastest. His services were highly sought after, and because of his schedule, he had to be booked months in advance. He traveled throughout New England and New York State to shear sheep, some 10,000 sheep annually, as well as alpacas and angora goats.

A freshly shorn purebred Southdown ewe, compliments of shearer Dave Hinman. He is part of a dwindling group of skilled laborers who serve farmers.

Hinman was fifty-seven years old, divorced, and had been shearing sheep full-time for nine years; he had learned to shear when he was a kid in Holderness, New Hampshire, and had been doing it in some capacity ever since. Even when he'd worked as a teacher full-time, he'd still managed to shear 5,000 sheep each year. He came shambling in, dressed in an old sweater and underneath that, a flannel shirt, and worn trousers and a baseball cap from which straggly dark hair peeped out. He had a compulsively garrulous manner, always joking, opinionated, a natural raconteur and performer.

"I don't drink, I don't smoke, I don't have any bad habits," he said. "Sheep are my children."

The sheep had been corralled in a pen inside the horse barn. The horses, Michael and Bobby, were outside in the snowy pasture. The reasons to shear the sheep now, in late winter, as opposed to waiting until it was warmer, or until they went out to pasture in May, were various. Over the past year, dirt and hay had accumulated within their thick, nearly impenetrable wool. If the wool was too matted and debris-filled, it would have less resale value. (Jennifer and Kyle would sell some of the wool to an organic spinnery in Putney, Vermont, and keep the rest to use as mulch in the gardens.) Because the ewes were three weeks away from lambing, shearing them would give Jennifer and Kyle a better idea of how close the ewes were to giving birth; a heavy wool coat masks a swelling belly. And it would be easier for a lamb to find a ewe's udders when they were not hidden by wool. Finally, Dave was simply too busy in the spring to fit them into his manic schedule.

He walked into the pen and selected a ewe. The other sheep, alarmed by the interloper, clattered over the cement floor to the window that looked out onto the pasture. They huddled together in a corner, baah-ing, looking away from him resolutely, as if ignoring him would make him go away. He maneuvered a sheep away from the flock in a mini-game of tag, with the ewe going one way and Dave the other, and then back again. When he had the ewe where he wanted her, he put his hand inside her mouth and guided her as gently as he could from the pen; pulling a sheep by its coat, when the wool is the object, is considered a kind of violation. He brought the sheep into the aisle that separated the horse stalls from the pen where the sheep were gathered.

Dave tips a ewe onto its rear end for a haircut as a Border Leicester lamb watches and waits her turn. The lamb was raised by a bottle because her 14-year-old mother didn't survive the birth. The ewe's three daughters, two granddaughters, and one great-granddaughter continue to live on the farm.

He flipped her onto her back, which rendered her more or less helpless. In this situation, sheep behave, he said, as if they have been caught by a predator, and are utterly submissive and paralyzed. Only the occasional sheep struggles by kicking its legs, but it is quickly and easily subdued. Turned on their backs, sitting up with their legs sticking out awkwardly from all that wool, they look incongruous and undignified and slightly humiliated by it all.

He began on the chest, around the armpits, making long, smooth shearing motions down the belly and around the stomach, removing the wool as a whole piece, as if he were slipping off an overcoat—which he was. "It's intimidating to shear sheep. The shears are very sharp," he said.

He worked quickly. Within five minutes, the ewe was completely shorn and her coat had been tossed by Jennifer onto a growing pile of wool that would, once all the sheep were shorn, reach nearly to the ceiling. Its bulk notwithstanding, the wool was lightweight and oily to the touch, because of the natural lanolin in the wool that acts as a water repellent and keeps the sheep dry and warm in bad weather.

Once shorn, the sheep go through an initial period of confusion; without the encasing, identifying cloud of wool, they lose their bearings. One denuded ewe was sniffing at the hindquarters of another denuded ewe, as a dog would upon first meeting another dog.

"A sheep can recognize two hundred different sheep," Jennifer said, as she watched the two ewes. "They recognize their appearance. When they're sheared, they get all confused."

As Dave worked, he and Jennifer gossiped about people in the sheep business. So-and-so's a jerk; so-and-so's a cheapskate; so-and-so is mean to his wife; so-and-so doesn't like another so-and-so. Unlike Kyle, who tended to keep his opinions about people close to the vest, the only visible or audible sign of his annoyance with someone a certain tightness in his expression and an uninflected terseness in his speech—"Yes," "No," "Okay"—Dave, like Jennifer, was vehement in his likes and dislikes, used the pitch of his voice to express outrage, humor, disdain, and skepticism. Both he and Jennifer had a penchant for raising their voice, not born of anger, but as a kind of comic exaggeration to make a point. He declaimed and she declaimed back.

Having disposed of George W. Bush (Nay) and Scott Ritter, former UNSCOM weapons inspector turned anti-war prophet (Yea), Dave was waxing lyrical about the various women with whom he was, not too seriously, infatuated. There was a beautiful woman over in this part of the state, and another beautiful woman down in that part of the state, and there was the woman whose house he used to call when he knew she wasn't there, just so he could hear her voice on the answering machine. I wonder if so-and-so would divorce her husband and marry me, he mused.

"You've got a girlfriend," Jennifer pointed out.

"She loves me. I don't know what her problem is," he sighed. They lived together on the farm near Walpole, and raised sheep there. The farm belonged to a wealthy couple who used it as a summer place; Dave and his girlfriend acted as caretakers. When the wealthy couple threw parties or offered chamber music recitals on the grounds, they asked Dave to put out the sheep to graze, to make the land look that much more bucolic, a country estate belonging to gentry. The owner had once come up to him with a very sober expression, saying he was very sorry to hear there had been a tragedy on the farm. Hinman mimed bafflement as he told the story, as to his mind nothing resembling tragedy had taken place. No one had died, the farm hadn't burned down, there was no obvious disaster. When he asked the man, who was not a native but a summer person, what he was talking about, the owner said he had heard that two lambs had died.

"Tragedy?! That's not a tragedy," Hinman said with a laugh. "That's farming."

As he sheared, the hooves of the sheep never touched the ground; in fact, the sheep hardly moved at all, which was a testament to his speed and skill. Jennifer admired the wool as she took it from him: "Beautiful."

Dave grunted. "Yeah, nice and crimpy." He complimented the sheep: "They look really good."

"They haven't gotten any grain."

"Yeah, Southdowns don't need it." He made a dismissive gesture.

"Do you want some help?" Jennifer askeds.

"NOOO, I don't want any help," he yodeled. He never lost control of the animal; each sheep was virtually immobilized until he released it, at which point it scrambled up gratefully. When Jennifer opened the barn door leading to the pasture, the ewe stared blankly for a moment through the door, where freedom beckoned, and then dashed outside, as if the devil himself were on her heels.

The flock was made up of a range of breeds: Southdowns, Romneys, Border-Leicester-Suffolk cross, Romney-Merino cross, Dorset Fins, and a few Montadales. The Southdowns predominated. They are a hardy breed, small in stature but endowed with broad heads that look a bit too big for their bodies and small ears that lie flat against their heads. They are all white (really a grayish white), tending toward a robust stockiness, considered to be an excellent meat lamb but not valued particularly for their wool. The Border-Leicester-Suffolk cross has a slimmer build and grayish wool and a narrow head and face, all black, and longer ears that stand up from the body, in contrast to the Southdown. The Southdowns tend to ignore humans, unless food is in the offing; the Border-Leicesters are curious and tame enough to smell at you if you draw near, and will not jump back if you put your hand out to them.

Jennifer and Kyle had begun, in 1998, with a foundation flock of sixteen sheep that they had acquired from the Billings Farm & Museum in Woodstock, directly across the road from the Marsh-Billings-Rockefeller National Historical Park. The farm manager had wanted to reduce the flock, and had sold them the sheep at a lower price. Jennifer and Kyle had lost three ewes out of sixteen the first year. Since that time, they had steadily increased the flock through breeding, and the occasional purchase: six Romney-Merinos from a breeder in Massachusetts, and four ewes from a dental hygienist in Vermont. And they were given two ewes by a man who couldn't keep up with trying to bottle-feed them in his bathtub during a family crisis.

The Fat Rooster ewes were bred by the two rams in August and September. After a six-month gestation period, ewes lambed in the spring, in March and April. Many lamb producers waited until their ewes were more than a year old to breed them, because of the body development. Animal pregnancy and birth, like human birth, can be fraught with dangerous complications, with risk to mother and offspring. This was particularly true for younger ewes who were less than a year in age; they might abort more frequently or suffer dystocia—a prolonged and difficult labor. They, or their lambs, might die as a result. Despite this inherent danger, Jennifer and Kyle bred ewes who were less than a year in age. Their rationale for doing so was twofold: one or both of them were always on hand to attend to lambing, should something go wrong, and, more to the point, breeding the ewes when they were younger meant getting their money's worth out of them. (Ewes have an average life expectancy of seven years.) More lambs, more money.

When the ewes began to lamb in earnest, as they did at the end of March and beginning of April, Jennifer and Kyle awoke every two hours during the night to check on them. This year, forty-one ewes, out of fifty-one, were pregnant. Ewes typically twin, and sometimes even give birth to triplets; ewes pregnant for the first time often give birth to just one lamb. Two of the ewes had already had to be killed, because of labors gone awry. It was too early to anticipate a streak of trouble, but it couldn't be counted out either. Of the five lambs born so far this winter, only one was still alive. This made Jennifer nervous that they were in for a run of bad luck, similar to the one they had had with the piglets. (Of fifty-eight piglets, twenty-three had died because of the cold—a nearly 35 percent loss, when 10 percent was the average. Thanks to warmer temperatures and cleaner pens, the piglets were now thriving.)

Some ewes will give birth unattended and without incident and know, instinctively, how to care for their lambs within moments after birth. Others have to be shown what to do. Jennifer sometimes acted as midwife, intervening when necessary, often going so far as to pull the lambs out of them, which entailed sticking her arm into the ewe's vagina as far as she could go, and feeling around—a head here, a leg there, two sets of legs entangled, a breech birth, or, in the worst case, aborted lambs, now decomposing, that had to be pulled out limb by limb. Once delivered of lambs, it was not unusual for the younger ewes to seem unprepared for the responsibilities of mothering.

In some instances, when ewes refused to nurse, they and their lambs had to be isolated within a separate pen so that the ewes did not physically shirk nursing, thereby depriving the lambs of the

A Polled Dorset lamb scampers through the pasture.

rich, fattening milk that they needed and wanted. And then there were the worst cases that ran into terrible, annihilating difficulty, who aborted their lambs and had to be killed, or who delivered lambs but had to be killed anyway because they had been torn apart, literally, by labor.

On March 19, the day before war began in Iraq, Kyle had called Jennifer at the Country Animal Hospital to tell her that one of the pregnant ewes was in terrible shape, and had prolapsed (a condition in which one or more organs, usually the uterus or rectum, slip forward and down, extruding from the body). "We're losing that ewe," he told her. "Her intestine is coming out."

"When I got there," Jennifer recalled, "her entire insides were coming out of her vagina—her stomach, her intestines. She was completely ripped open. I'd never seen a description of that in the books. I slit her throat to kill her, to put her out of her misery. Both lambs were dead. They were rams, and beautiful. I was a total basket case. She was in so much pain and shock. I can't use guns to shoot sick animals. Gunshots are worse than cutting the throat. Before I went to work that morning I saw the look of death in her eyes. I should have just done it then, but I didn't want to do it. I was all dressed up for work."

She mused further on the intimate connection between life and death that she sees on the farm, which, in some way, seems a microcosm of the world around her. "Maybe if people knew that intensity, we wouldn't be bombing the hell out of Iraq."

Kathleen O'Boyle had been delegated lamb duty. She would take the midnight shift; Jennifer, the two A.M. shift; Kyle, the four A.M. shift. Kathleen had distinctively Irish coloring: blue eyes and glossy, blue-black hair that she kept cut close to her head, and deep dimples. She was in her twenties, from Massachusetts. She had a sturdy build that was complemented by a sturdy, easygoing, unflappable manner, and a quiet, self-contained detachment. She was rarely seen without a pair of jeans and a hooded sweatshirt. Like Russ, she was good with Brad, would play with him whenever he asked her to, and he sought her out.

At just before midnight, April 1, she pulled herself from bed, still fully dressed. She was sleeping in Brad's room; Brad slept in his parents' room, which had as much to do with a bout of night terror he'd been experiencing as it had to do with finding room for Kathleen. She padded quietly down the stairs. Putting on a jacket and mud boots, she tromped over the thick ice to the barn, swung open the big barn door, walked through the haymow, and then down the stairs into the livestock pens. She turned on a light and looked over the ewes carefully, most of which sleepily ignored her. There was no sign that any of the ewes were in labor, or close to it. The barn was unusually silent; even the sows were still, except for the occasional gusty snore. For whatever reason, the ewes had not been giving birth at night, as they had the previous year, but in the morning, between ten A.M. and noon.

There was one lamb inside the house, sleeping in a cardboard box by the woodstove. His mother, as sometimes happens, had refused to nurse him; each spring, some lambs were rejected by their mothers. Jennifer had brought him in and was feeding him by hand, with a bottle. He was weak, malnourished, listless, his skin hanging in folds, indifferent to Jennifer's touch. His chances of survival were slim, and indeed, he would later die.

The next morning, a Wednesday, Jennifer was not there. She had gone to work at Country Animal Hospital: she was on duty Mondays, Wednesdays, Fridays, and sometimes Saturday mornings; off on Sundays, Tuesdays, and Thursdays. Kyle, who, as a rule, worked Mondays and Tuesdays in Woodstock, was alternating between pruning the apple trees below the house, looking in on the lambs and taking the truck to pick up ironwood logs that he would use as hosts for his shiitake mushrooms. Today he barely got back, having gotten stuck.

This was mud season in the north country, March and April. Mud season is not hard winter, but it is not yet spring. It is winter grudgingly ceding ground, inch by unprepossessing inch. In mud season, the back roads are deeply rutted quagmires, nearly impassable because of the frequency of cars and trucks on narrow, twisting roads that weren't designed to accommodate the heavier traffic that has resulted from the population growth in the region.

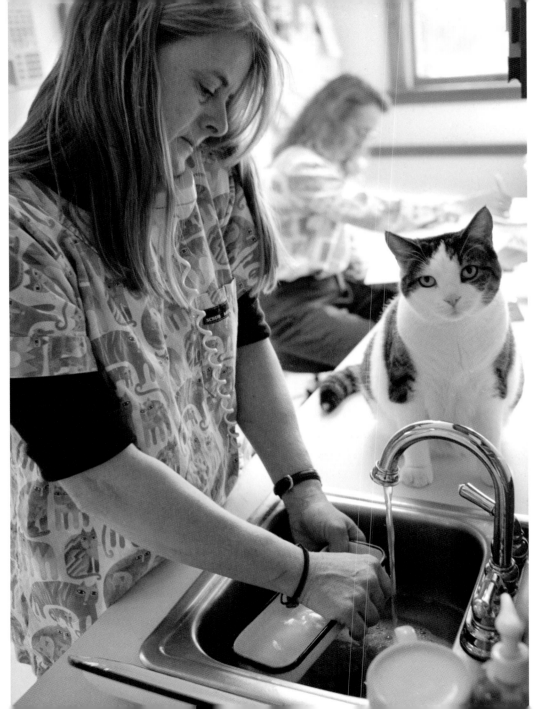

Mr. Crunchy the cat watches a visitor while Jennifer makes a phone call and does cold sterilization of equipment at Country Animal Hospital in Bethel, Vermont. Jennifer works at the hospital part-time. Owner Dr. Laurie Barcelow is in the background.

The lowering clouds threatened rain or snow or both. The air had a raw, mercurial quality—fresh and hopeful one minute, harsh and cutting the next, as likely to lash your face with sleet or wind as to caress it. There was not much sun—more light in the morning and in the afternoon, but it was a soupy, dingy light. The snow, after months on the ground, was dirty and unappealing. The plow trucks, over the course of winter, had shifted the snow into great piles that were now almost black with silt and grit. The paved roads were pitted and scarred and seamed by the frosts of winter, the macadam crumbling and worn away. The landscape was drab, monochromatic; although if you looked closely, there were subtle variations of ochre, charcoal, amber, and olive threaded through what looked like uniform grayness.

The animals were tired of being penned up and penned in. So were the humans. It was too early to see the advance of spring—no grass, no shoots, no buds—just bare clods of earth and matted grass and decomposing leaves from the previous autumn. The ice on the brooks and streams and rivers was softening and breaking up; one good warm spell, and freshets would gush forth untrammeled, and there would be talk of flooding.

Logic told you that spring was coming, that the robins would return, that the blackbirds would return, that by the end of April there would be a short stubble of new grass peeping up through all the brown, and that a few weeks after that, it would be an improbably brilliant, almost cartoonish green everywhere you looked. But mud season defied logic, and turned expectation on its head. It dragged on and on and on.

There was life stirring within, if not without. Nearly all the ewes were at or near term. Their movements were ungainly, their bodies uncomfortably swollen. They sank down into the hay with visible relief, and chewed their cud and stared myopically into space. From the side or the back, they looked like barges ferrying heavy cargo, moving slowly downriver. They were so characteristically stoic that you had to look for very subtle clues that they were at the beginning of labor—if a ewe were not chewing her cud, for example, or if her ears were pinned back against her head, both of which indicated distress. One ewe began to shift about restlessly, moving from one spot to another, a sign that she was in or near labor.

By the end of lambing season, a few weeks away, the sheep population would have nearly doubled. Forty-one ewes would birth sixty-seven lambs in all, of which thirteen would die. The pens would hold a large, woolly, teeming, baaah-ing mass of sheep, as well as the two rams which, between them, had sired all those lambs (and which, if you got too close to mothers and offspring, would stare fixedly before lowering their heads and stomping their feet in warning). By the end of the year, taking both replacement stock and culling into account, there would be sixty-one ewes and two rams.

Although it was relatively early in the lambing season, some ewes had already given birth; their lambs nuzzled against them, or butted their heads against their mothers' teats, to bring down milk. Other ewes pawed gently at their offspring to remind them to nurse. One newborn sprang giddily in the hay; its movements were not yet fully coordinated, as if it were a marionette whose strings were being jerked upward by an unskilled puppeteer. The lambs would run at top speed from one end of the pen to the other, leapfrogging over their mothers in what Kyle called a lamb run, an exuberant display of new life that he found endearing and would take time out from chores to watch.

The ewe that had been moving around uneasily suddenly looked behind her in bewilderment. A large balloon-like bag was emerging from her vagina; filled with blood and water, it hung to the floor, intact. She was unnerved by it, kept twisting her neck around to look at it, tried to kick at it with her back hooves, as if she had little or no idea what was happening to her and was frightened by it. A small white head, still covered by the sac, emerged from her vagina. The outline of the head, as seen through the opaque membrane, was blurred.

The ewe stood rooted to the spot for what seemed like an eternity without delivering. As if she had given up, she then made a nest, scraping at the hay with her hooves, and lay down awkwardly. The lamb appeared to be stuck. Within minutes, the ewe's neck arched up and back and her eyes rolled back in her head. She grunted, a visceral sound, and her back end started to rise, as if of its

Jennifer, tending to a ewe on the verge of lambing, greets a lamb that's a "milk stealer"—it goes to ewes that aren't its mother to nurse. In 2003, Fat Rooster Farm had sixty-seven lambs, a number that grows every year.

own accord. You could see that she was now experiencing severe contractions, her belly appeared to be flexing. She involuntarily lifted one of her back legs off the ground, and moaned; a prolonged moan, during which she pressed her face hard up against the back wall of the pen, as if that brought relief. There was a sudden heave, and the sac was propelled forward. The lamb was born, but the ewe seemed confused by all that had happened, and did not attend to it immediately. The lamb lay in a heap, not moving. The ewe sniffed at it and looked away dully.

Kyle came in to check on the ewes. He saw that this ewe had had one lamb, but that the lamb was not moving. He wiped its nose and mouth clean, then picked it up by its legs and swung it back and forth as if it were a pendulum, to animate it. It fluttered weakly, still alive. The ewe seemed to take her cue from Kyle. She began to lick away the afterbirth, and did so obsessively, as if comforted by the repetitiveness of the motion. This kind of robotic licking was also a stimulant to further contractions; in all likelihood, there was another lamb inside her.

Kyle left again, momentarily. She paused, left off licking. Again she went through the neck arching, the eye rolling, the back lifting off the ground. A second lamb was extruded; it was born more quickly than the first, and its sac broke immediately. Again, the ewe seemed at a loss as to what to do with it. The newborn lay in a heap, while she concentrated on the first lamb. Another ewe came over, concerned, and acted as a mother to the second lamb, scouring away the blood and ordure with her tongue. This second lamb moved, tried to get up, and did so fairly quickly.

Kyle returned and looked at the first lamb, which had now lain utterly motionless against the wall for ten or fifteen minutes, its mother's attempts to rouse it notwithstanding. "It doesn't look good," he said, and climbed back into the pen. He again tried swinging it back and forth vigorously, then lay it down and tried resuscitation, getting down on his hands and knees and blowing puffs of air into the lamb's mouth. It was a futile effort.

"It's gone," he said. He stared at it for an instant, moved it aside, and then without hesitating, took the second, neglected lamb and put it firmly underneath the ewe's nose. As if nothing had happened, and there had never been another lamb, she continued the cleaning process.

In hindsight, it was obvious that the first lamb had been dying. For the half hour that it had lived, it had done so in a kind of shadow world, not quite living, not quite dead. It had never had the look of life to it—vigorous, unabashed, hopeful life. It had never tried to stand, but lay inert, its fragility apparent, as had been its mother's confusion and disorientation.

"It was probably for the best," was Jennifer's later dispassionate analysis. "That ewe wouldn't have known how to take care of two lambs."

The dead lamb was disposed of in the gutter cleaner, still warm. While Kyle was examining it, wondering at its demise—which he attributed to, in all likelihood, the fact that its mother did not immediately clean away afterbirth that clung to its nose and mouth, leaving it unable to breathe— his attention was caught by something in the sheep pen opposite. A movement. A flash of something that wasn't there a minute ago. Without fuss or noise or trouble, another ewe had given birth.

It was snowing the next morning, the kind of wet, heavy, lumpish snow that often comes in early April. Jennifer was keeping a close watch on the ewes. There was one that might be in trouble. Jennifer had had her eye on this ewe since before dawn, when she had seemed to be in labor. Jennifer had already lost another ewe two days earlier, on Tuesday, March 31—ewe #37, whose throat she had cut when it became clear that she would not survive, although her two lambs had. They had joined the lamb living in the cardboard box inside the house.

"She had a big, huge lamb and then a little lamb. I think she ripped something inside," Jennifer said later. "Her whole back end was swollen. I came down at three in the morning and she was in shock. Her ears were down; she had this pained look on her face. I had no compunction over killing that ewe. I knew it would cost me more to save her. She was suffering."

It was now six or seven hours after dawn, and the problem ewe still appeared to be laboring. In human terms, a six- or seven-hour labor would be considered reasonable, even short; in ewe terms, this sort of prolonged labor is often the sign of a birth process that has gone wrong. The ewe was groaning, arching her back. With one contraction, her lips retracted, baring her teeth in a rictus of pain. She had moved over near a wall to lie down, and as if by tacit agreement, the other sheep, who usually crowded close, had respectfully receded, leaving an arc of empty space around her. The dead lamb from the day before still lay in the gutter, its fleece now littered with hay and manure.

The ewe shuddered. A movement of her flanks looked, for a moment, like wind rippling across water. A lamb slid out of her and onto the hay. Immediately, a nearby ewe sidled over to sniff curiously and protectively at it, as did her two lambs, one white, one black. The visiting ewe started to lick the newborn lamb's legs clean, a show of mothering that was a kind of displacement behavior, rather than a gesture of pure altruism. The laboring ewe joined in, compulsively lapping at the lamb's face to remove the afterbirth, until another set of hard contractions forced her to brace herself. She groaned, a sound torn from her involuntarily. A second bag popped out, burst, and this second lamb shook itself briskly and raised its head. Steam rose from its skin. Despite Jennifer's fears, both lambs had been born easily, looked to be strong and alert, and their mother had recovered her equilibrium fairly quickly.

Back in the house, Jennifer tried bottle-feeding the three lambs, using not mother's milk, but milk replacer. She judged it too much trouble to milk ewes directly, given that these house lambs would probably die anyway. Only one of them, a lamb that Brad had decided to call Presto, was relatively lively. The other two were having trouble; their breathing had the labored, mechanical sound of a suction pump, and when she tried to get them to stand on their own, they collapsed into a heap of tangled limbs. Even their feeding seemed weak; no urgent sucking of the nipple but a timid, inexpert fumbling, in which more milk ended up on the floor than in their mouths. Without the guidance, and the physical presence and reassurance of their mothers, they had no idea how to behave or to survive or adapt. They sank into a sleep that looked closer to deterioration than rejuvenation.

If they lived, they would be smaller, unlikely to breed, and never truly part of the flock, but more attuned to the humans, following them around like goats would. This made them, in Kyle's eyes, a liability, because they would always require more attention and would not fulfill their biological imperative, which was to breed. Jennifer shrugged it off as something that just happened, and you tried to save them if you could, although she would also be quick to ship them to slaughter come summer if they proved troublesome.

Now that Jennifer and Kyle were in the thick of lambing, with lambs coming every day, the sap had, of course, finally decided to run with a vengeance. Agnes and George were pressing them to get out and collect, so that they could salvage what was left of the sugaring season. One dilemma had resolved itself on its own; the problem ewe had lambed without trouble. That left the problem of the sap. Jennifer decided she needed to enlist additional help because Kathleen was in the greenhouse, transplanting, and Kyle was up in Randolph, driving the school bus. He was paid $25 for a two-hour run; not a magnificent salary, but it added up if he did it a few times a week, and this was the farm's downtime, economically and logistically. The man who usually drove the bus, and had been doing so for twenty-four years, had been fired. (The local newspaper would report that he had been accused of making sexual advances toward two boys who rode on his route, a crime for which he was later convicted.)

Jennifer flung herself into a chair and called Russ. The conversation went like this: "Hey Russ, what are you doing this afternoon? . . . Where did it bite you? And *why* did it bite you? . . . At least it wasn't rabid. Well, I guess you aren't up for gathering, then." Russ had been trying to free a cat from a rat trap and had been bitten for his pains.

Brad was underfoot, and after she fed him lunch, Jennifer would have to find a way to deal with him. For safety and sanity's sake, it was not advisable that he ride on the tractor with her when she went out to collect sap. She then called Debbie.

"Hi, Debbie—can I drop Brad off from one-thirty to four? I can't have him and drive the tractor at the same time." She arranged with a friend named Gene Craft—the husband of her close friend, Laura, with whom she worked at Country Animal Hospital—to help her collect the sap.

Kyle carries a pump and pipe to empty a load of sap into a holding tank. The sugarhouse's stainless-steel holding tank used to be the bulk tank used for milking when the previous owners of the farm produced dairy.

Jennifer, left, and apprentice Meg Wedding move washed salad greens to the "spinner"—a dedicated clothes washing machine—in preparation for the weekly Lebanon Farmers' Market and CSA subscribers' pickup.

They had hired three apprentices for the summer: a couple coming from a stint on a dairy farm in Potsdam, New York; and a young guy, Bill Fielding, who lived in Hartland, Vermont, and had interviewed for the job in February. It was something of a surprise that Bill Fielding had taken the job. When he was interviewed in February—a tall, dark-haired, genial twenty-four-year-old with a surprisingly stentorian voice for one so young—he had seemed unsure of himself as he shifted from foot to foot in the driveway, hands thrust into jeans pockets. He had talked about going west with a friend— the Jack Kerouac, *On The Road,* American rite-of-passage of hopping in a junk car and driving to California. He'd talked of wanting to earn more money than the farm could pay, maybe as a bartender. He'd talked about the meaning of life and death, and philosophy, which he had been studying at UVM before he decided to take time off, and whether he wanted to continue college or whether it would be a better education to travel the world. He'd talked about farming and the land and growing up in Vermont. He'd talked about working the previous fall on a farm down in Cornish, New Hampshire, where he'd killed turkeys for Thanksgiving, which was, he said, quite an experience, an eye-opener.

He'd talked, in short, about everything but a firm commitment to the job. He did not want to be pinned down, and made vague-sounding promises about being in touch. Jennifer was not unsympathetic; he reminded her of herself at that age, when she felt all at sea and pulled in six different directions at once, searching for something that made sense to her and wasn't all abstraction. She knew well that feeling of being the outcast, the odd man out.

Earlier that morning, before Bill's arrival, she had interviewed another candidate, a young woman who had written a very thorough, cogent application on why she wanted to work for them, and what she would bring to the job. Sadie MacKilop was majoring in environmental studies and anthropology at UVM and had grown up in Middlebury, but was now living just down the road in Sharon. She and Jennifer had sat at the table, hopeful sunlight washing over them, while Jennifer reviewed again her application. Sadie had said all the right things and seemed eager to please and to learn and to work hard. She was a vegetarian but had said, when pressed, that no, working around animals that would be slaughtered would not disturb her or compromise her vegetarianism.

She had also put in applications to other farming operations that paid a higher wage, and Jennifer was doubtful that she would, in the end, sign up with Fat Rooster. She was the kind of candidate who would be snatched up by any one of the farms in the area; she seemed focused and smart, and knew something about farming. She had said she'd get back to them, but that was one of those things that people say when they're hedging their bets, and they didn't hear back from her. Later that spring, Bill Fielding had called back and said that he was interested in taking the job.

The couple who had signed on, Ben Canonica and Meg Wedding, were in their mid-twenties, which made them older than the usual run of apprentice, and they were very experienced, having worked at farms and ranches throughout the United States. They also were asking a larger salary: $130 each, weekly. That was well below the minimum wage of $5.50 per hour, when you broke it down into a day's labor—$21.66, or $2.70 hourly for an eight-hour day, and farming was rarely an eight-hour day—but it was more than Kyle and Jennifer had ever paid an apprentice, for whom wages were supposed to be in the nature of an honorarium, a garnish to the education and experience they would receive in farming, as well as the room and board and meals furnished.

The relatively high salaries notwithstanding, Kyle and Jennifer had decided, in light of the previous year's experience—when some of the apprentices were haphazard in their attentions to their work—that they preferred to pay the higher wage for more reliable employees.

On March 12, Jennifer, along with some other poultry producers, had driven up to Montpelier to attend a hearing held by the House Agricultural Committee. It was the second of three trips she would make to Montpelier during the year in the hopes of persuading the Agency of Agriculture to grant small producers like Fat Rooster Farm a federal exemption that would allow them to sell poultry at farmers' markets, which they currently were not allowed to do.

The exemption, which fell under the Poultry Products Inspection Act, as overseen by the Food and Drug Administration, permitted producers of not more than 1,000 birds per calendar year to sell to consumers, restaurants, hotels, and boardinghouses, as long as those sales did not cross any state or territorial lines, the product was labeled with the name and address of the producer, and was "sound, clean, and fit for human food."

There was, of course, a market for free-range, locally produced chicken, but as it stood, it was illegal for them to sell or transport poultry off the farm in Vermont. (They were allowed to sell chicken at farmers' markets in New Hampshire, because New Hampshire approached the regulation differently.)

Jennifer and Kyle had already run into problems when, in 2001, they had sold chickens to a local food cooperative. The chickens had been slaughtered by a man who was, they believed, federally and state certified in Vermont; that turned out not to be the case. An inspector for the state told them additionally that they did not have a license to sell to restaurants, and grocery stores like the cooperative fell into that category.

Jennifer had assumed that Vermont granted the federal exemption allowing producers of not more than 1,000 birds to sell uninspected product under certain conditions, but that was not so. The federal exemption was granted at the discretion of the states, and in Vermont, that power fell not to the legislature, but to the Vermont Agency of Agriculture, Food & Markets, specifically, the Commissioner of Agriculture. (In May 2003, Governor Jim Douglas would sign a bill that elevated the Commissioner of Agriculture to a cabinet position; henceforth, the Commissioner became Secretary of Agriculture.)

After the inspector informed her that they were in violation of state law, Jennifer said, "Okay, what do I have to do to sell my chickens?" The answer was, they had to be federally or state certified. Which, from her point of view, was easier said than done.

The state and federal regulations concerning what meats and poultry can be sold on or off the farm are, to say the least, byzantine, and awash in contradiction. If you are a producer of meat or poultry—in excess of 1,000 animals or birds—and you want to sell to retail outlets, the regulations are straightforward; the meat or poultry must be federally or state inspected, which means, in effect, it must be slaughtered at a federally or state-inspected facility.

Few farms in Vermont are big enough to have those kinds of facilities on-site, or big enough, in terms of sales, to command the visit of a certifying inspector. Logistically speaking, in a small state like Vermont, there are thousands of farms but only a small number of inspectors, state or federal. It is up to the farmer to seek out federally or state-certified operations if he or she wants to sell their meat or poultry at retail. (Federal certification permits interstate commerce; state certification, only within the state.)

This has been made more difficult by the fact that there are now only ten red-meat slaughterhouses in Vermont, six of them federally certified, four of them state certified. There are three poultry slaughterhouses in the state, two federally certified, and one state certified. Of the two federally certified poultry processors, one will not process other people's poultry, and the other will do so only in large numbers. The third was state certified, but, from Jennifer's viewpoint, charged a high processing fee, in excess of $3 per bird, and was nearly two hours away. This made it impractical for a smaller producer like Fat Rooster Farm, raising in the neighborhood of a hundred meat birds annually, and slaughtering in batches of twenty-five to fifty, to enlist the services of these larger poultry processors.

There simply were not enough processors, of either meat or poultry, to serve the needs of the farmers in the state, and the ones that were in operation had suffered financial setbacks that threatened to put them out of business. The decrease in the number of slaughterhouses can be attributed in part to the effect of vertical integration by agribusiness—the larger factory farms raise and slaughter livestock or poultry on-site, rather than contracting with independent producers and processors at steps along the way.

There are, however, exemptions to the regulations concerning inspection within the federal statutes—the Federal Meat Inspection Act and the Poultry Products Inspection Act—that pertain to the production and processing of livestock. These exemptions fall under the category of personal and custom slaughtering; that is, if you raise a steer, chicken, turkey, or pig for your own use on the farm, you need not have it inspected. If you then serve that steer, chicken, turkey, or pig to your family or to guests or to employees, you are still exempt from regulation.

A Fat Rooster Farm cut of veal is displayed for sale in a cooler at the Norwich Farmers' Market.

The minute you wade into the area of retail sales to the public, the regulations become more complex and stringent. This is of necessity; there is a standard and perception of public health to maintain. Jennifer and Kyle sent their sheep and pigs in 2003 to federally inspected slaughterhouses, in Rutland and Ferrisburg, Vermont. And although they couldn't sell poultry to retail outlets, they were permitted to sell poultry to consumers at the farm; the state saw that as a case of caveat emptor, or buyer beware. The consumer assumes any risk for buying poultry that has not been inspected, presumably because the consumer either has conducted his or her own inspection of the farm, and is satisfied that basic safety standards are being met, or knows and trusts the supplier.

The loopholes and exemptions are, from the point of view of the small farmer, often bewildering: they are permitted to sell their veal or baby beef off the farm directly to consumers, as long as it is sold prior to slaughter, or what is termed "on the hoof," and is sold by the whole or the half. In other words, you might buy a whole calf or part of a calf and wait for it to grow to the age when it can be slaughtered, and then collect your share of that calf—a transaction that takes place entirely without government intervention or inspection, and is perfectly legal, as long as it falls into the category of what the USDA calls "custom slaughter" for the household, its employees, or guests. If, however, it is going to be sold to the consumer in sections—steaks, ribs, loin, chops—it must be federally inspected.

Frustrated by what she perceived as a bureaucratic indifference to the small poultry producers, Jennifer, along with some other farmers, had independently helped to arrange a preliminary meeting with the eleven-member house agricultural committee, a tactic that she later acknowledged did not go over well with some of the people in the Agency of Agriculture, who felt that she had gone over their heads.

In any case, the hearing yielded little in the way of immediate action, except establishing that there was need for further hearings. Jennifer was there. Tim Sanford and Suzanne Long from Luna Bleu farm in Royalton were there. Carl Cushing, the director of Food Safety and Customer Assurance, was there. Marian White, an agricultural policy analyst for the Agency of Agriculture, was there. Eight members of the committee were on hand. The meeting was held in the morning in a cramped room on the second floor of the statehouse in Montpelier—folding chairs and stale air and discreet boredom. Jennifer sat in a far corner of the room, with a sheaf of papers on her lap. Her hair was down, not in a ponytail as was usually the case, and she had put on a pair of dressy white linen pants. Her brow was furrowed and the look on her face was tense, as if she were prepared for battle.

"I don't think we can legislate; we can facilitate," said Ruth Towne, chair of the committee, a dairy farmer herself, and a Republican from Berlin, a town sandwiched between Montpelier and Barre immediately to the south. Towne was addressed as Madame Chairwoman. In her seventies, with a phlegmatic manner, she wore a red suit and glasses and looked like your grade school grammar teacher.

The discussion quickly boiled down to this: that there were going to be, this coming summer, only two available processors of poultry in the state; that access to those processors was limited; and that perhaps a way out of the stalemate was to raise the money for a mobile slaughter unit that could travel from farm to farm. Of course, such a unit would be expensive, and it would have to meet the state criteria, and an inspector would have to be present to certify the poultry.

"It doesn't matter who does the slaughtering, as long as an inspector is there?" Tim Sanford asked.

"Yes, there needs to be a qualified person there," Carl Cushing answered.

"We're trying to build momentum and will for an outcome that everyone can agree on," murmured Bill Botzow, a Democrat from Pownal, in the southern part of the state. Unlike some of the other committee members, themselves farmers, he wore a suit and tie, and had the look of an academic. "We're building a consensus of where you think you're going."

Jennifer jumped in. A cooperative mobile slaughter unit, she argued, if it was to happen, was at least two years away, and would cost in the neighborhood of $10,000. It would involve seeking funds, and grant writing, which was a long process. In the meantime, she said, "it is in this legislature's power to grant us those federal exemptions. We could use your help to enforce that federal legislation." Her tone was earnest but with a hint of strain.

"The exemption is at the discretion of the commissioner. Do we want to do that?" Carl Cushing asked. "If the exemption is removed, small farmers may have competition they don't want. The exemption keeps a distinction between inspected and non-inspected meat. Until we can get something together, I'm hesitant to remove the exemption."

Suzanne Long chimed in. "The exemption would enable smaller growers to develop markets in a smaller way. If we're putting time and money into a larger effort, it would be nice to start making those connections now. It would help us to develop a business plan."

More complications arose. Because the exemption permitted sales to hotels, restaurants, and institutions (HRIs), that became the province not of the Agency of Agriculture, but of the Department of Health. So now you had two different departments involved in one issue, which had the potential to muddy the waters. "It's the policing of it that's the problem," Carl Cushing said. There were seven inspectors in the state, which was not sufficient, he felt, to conduct spot checks on all the farmers' markets that might open up if the state granted the exemption. "We deal on a complaint-driven basis, but otherwise we don't inspect markets," he added.

"What would it take for us to be certified in the short term?" Tim Long asked.

"I can't address HRIs," Cushing said.

"What about retail? Farmers' markets?" Long pressed.

"As far as certifying farms goes, we can provide the training," Cushing allowed. "I'd be willing to sit down with you folks to go over certification."

And that was that. The hearing had gone over its allotted time limit, with no apparent substantial movement forward, or at least not one that made Jennifer happy. The only way out of the impasse, it seemed, was to pool resources for the eventual purchase of a mobile slaughter unit, or for interested farmers to look into converting a part of their operations into a certified slaughter facility—all of which would take time to bring about, if it could be brought about at all. The exemption was not going to be allowed; it was not couched in such direct language, but that was the obvious subtext.

The issue, Cushing and other members of the department maintained in a second meeting at the Agency of Agriculture on April 15, was one of consistency and safety. If the exemptions were allowed, how could they police all the markets? Sure, the producers who had bothered to come up to Montpelier were operating on good faith, and ran safe, clean shops, but what about the ones who didn't? What then? The people who worked for the Agency of Agriculture had seen such operations and they weren't pretty, and there were, they averred, more of those kinds of operations out there than the farmers here today could imagine, or would admit.

"We see a lot of people who don't do it right," Dallas Meek, then-state veterinarian, interjected.

And, further, the federal exemption was, from the state's point of view, immaterial. "Federal law doesn't apply. We regulate in-state commerce," said Curt Stasheski, section chief of the Meat Inspection Section. While the exemption was written into the Poultry Products Inspection Act, the USDA did not regulate retail commerce at the local level; the state did. Slapping on an identifying label, so that if there were any tainted product it could be traced back to its point of origin—one form of quality control advocated by the farmers who had attended the meetings—was not sufficient from the perspective of the inspectors. Anything sold to retail outlets in the state of Vermont had to be federally or state certified, period. (Game birds like pheasant or quail fell into a gray zone; because they were not covered by the Poultry Products Inspection Act, they were not regulated. Because they were not regulated, it was possible to sell them at farmers' market and not be in contravention of law.)

After the meeting, Jennifer went back to the farm discouraged and skeptical that there would be any movement forward on the issue. She felt she had been defeated; that no matter what was said or proposed, they would raise new objections; that they kept raising the bar. Further, the farm had now been cited twice for violations; the second time when they sold chicken to a man who in turn sold them, as part of prepared meals, to the same local cooperative where they had been caught out the first time. Jennifer and Kyle felt, rightly or wrongly, that they were under closer scrutiny, and they were not anxious to be perceived as perpetual agitators; their farm depended on maintaining civil relations with the state, the consumers, their neighbors. Too many violations, and the state

Pica Pica, an eight-year-old Plymouth Rock rooster and Fat Rooster Farm's mascot, roams the barn while Kyle does the chores. Due to its age and pecking order amongst the flock, the rooster is not kept in the coop.

would yank their retail license, which permitted them to sell at farmers' markets; they could not afford to lose that venue. They would let the matter drop, although it irked Jennifer, who had a combative streak, to do so.

What she wanted, in an ideal world, was for the state to recognize the federal exemption. In the world of the here and now, this was unlikely to happen, given the state's argument that it was a matter of public health. In the meantime, Jennifer and Kyle had worked out a new arrangement with Luna Bleu, seven miles away. Tim and Suzanne had the equipment to slaughter and to dress chickens and ducks; in return for letting them use it, Jennifer and Kyle would raise piglets for them. This would not change the status of their chickens, from the state's perspective, however. It was still uninspected poultry, and forbidden for sale at market.

In the summer of 2003, the Agency of Agriculture, Food & Markets did reach a compromise. A new protocol was established wherein small farmers would be allowed to sell uninspected poultry at farmers' markets, if they met certain conditions. The poultry had to be their own; they could not slaughter other farmers' poultry. The farmer had to provide potable water, hot and cold, for slaughtering and dressing. The facility had to comprise a "dirty" room and a "clean" room: the former for killing, scalding, and plucking the birds; the latter for eviscerating, washing, chilling, and packaging the birds. The walls and floors had to be impervious to water, not porous—that is, they had to be glass board or concrete. There had to be ample equipment on hand to process the birds according to specification. If a farm could meet those criteria, it was not necessary for an inspector to be on-site each time birds were slaughtered. Records would have to be kept, and the facility would have to be inspected once or twice a year.

To that end, Jennifer and Kyle invited some inspectors to come out to look at the farm, to see if there was a way they could remodel their milk house to meet the criteria. They determined that it would take between $8,000 and $10,000 to outfit it properly, an investment they could not afford and that seemed far out of proportion to their output, which was 100 meat birds annually. (In 2004, they would find an experienced processor with a mobile-slaughter unit who would come to the farm at relatively low cost. Because he was not certified by the state, however, Jennifer and Kyle still could not sell their chickens at market; a source of continuing frustration to them.)

A gusty, warm, and humid wind was blowing up from the south, the kind of wind that feels more like Indian summer than spring. It was the middle of April. The fields were not yet plowed. It was too early; the ground was too wet and spongy to drive over it with a tractor, and Kyle would have to wait until it dried out. In the meantime, to occupy himself, Kyle had erected two workbenches out in the yard, next to the barn, near his wood shop. On the ground was a stack of about twenty logs, which he intended to inoculate with shiitake mushroom spawn that he'd bought from a supplier in West Virginia. They came in little white thimbles, which he would insert into the logs, one thimble per drilled hole. A buyer for a co-op had told him fresh mushrooms would sell well. There were so few logs it wouldn't be worth much, but he wanted to see if it would work. Kyle was not an ardent mycologist, but as an ecologist at Marsh-Billings-Rockefeller National Historical Park, he was involved in various aspects of silviculture, which, perforce, brought him into contact with the flora and fauna found there, mushrooms included. Hundreds of edible and inedible species grew on dead logs and tree trunks and stumps, or near their base, and sprang up almost overnight when there was heavy rain.

A young woman, Randy Shafer, from Croton-on-Hudson, New York, was staying with them. She had been looking for a farm to work on for a week. She was just one of the numerous younger apprentices who would pass through Fat Rooster Farm from year to year, anywhere from two to four apprentices. This revolving workforce comes to the farm through various organizations, like Appropriate Technological Transfer for Rural Areas (ATTRA), that act as brokers for farms needing inexpensive, seasonal labor, and for young people who are interested in getting experience working on farms.

Randy would help Kyle inoculate the logs. Like nearly everything Kyle did, how he went about executing this particular task spoke to order, although he tended to shrug off his endeavors as so much mucking about, crazy experiments that might or might not bear fruit, a studied casualness that underplayed the dogged attention he gave them. The logs were neatly stacked, the workbenches placed just so; there was a plan of attack. The thimbles looked like medicinal capsules, and they had a styrofoam cap that sealed in moisture, on which mushrooms thrived; they wouldn't grow without it. The logs were roughly all the same size; the larger the log, the more moisture it retained, although it also required more spawn.

Kyle liked working with wood and, whenever he could, used wood that came from their own property. Today, he would inoculate oak, ironwood, and cherry logs; oak was the best habitat for growing mushrooms, ironwood the second best. The cherry he happened to have on hand. It had cost him $25 to buy the spawn, but the real expense would come in the labor of inoculating and moving the logs.

Sugaring season was finished, thank God. Everyone had gotten sick of it, and tempers had been testy—George and Agnes wanted Jennifer and Kyle to do things their way, in their time, and Jennifer and Kyle didn't always necessarily want to comply. "The collecting and boiling of sap is done," Kyle explained. "I'm just in denial today. A lot of cleaning of pipeline needs to be done. We finally got over two hundred gallons; last year we had two hundred seventy-six gallons. All the syrup was dark or darker. A few people made fancy. It was an average year, now that it's over, but it was a struggle."

Kyle studied the logs thoughtfully. He began boring equidistant holes, in a diamond pattern, up and down the log. He bored eight rows of holes in all, four inches apart. They would use the whole log: "The reason to do the whole log is to make sure only shiitake come in," Kyle explained. "Different fungi use different parts of the tree." As he drilled holes, he pulled a pencil out from behind his ear and used its point to clear the holes of wood shavings.

Once he and Randy had finished, he would situate the logs out in the woods: one end of the log on the ground, the other end off. "If you're really into production, you use sprinklers," he told Randy. "I'll do seat-of-the-pants, water them once in a while. You can force production by immersing them in cold water. It mimics a big rainstorm, which is good for mushrooms."

If the shiitake experiment was successful—and the results wouldn't be known until May or June of 2004, because it takes a year for the mushrooms to spawn and then fruit—he would try to

Labels are attached to maple syrup containers. Depending upon its color, the syrup is graded to a standard set by the state of Vermont.

increase production. As he worked, he, consciously or not, gave Randy a natural history lesson of sorts. "The mycelium grows in the log, about an inch a week. It's the equivalent of the roots of the fungus; the mushroom is the fruiting part." He then began inoculating an ironwood log, which led effortlessly into a brief disquisition on that particular wood. "Ironwood is a species not respected by foresters and loggers because it doesn't grow into good timber. It's a nice understory tree, but it doesn't convert into lumber well. It's dense and has some sapwood. Mushrooms need sapwood. George Spaulding calls it lever wood. Ironwood: the name's a hint, too. I roughed out a couple of bowls this winter with ironwood, and it worked out well. It's a white wood; not a very interesting grain. Ironwood has little seeds that can be good wildlife food."

The next morning, he put the logs in his pickup truck, and drove it down to a point in the pasture where it was easier to get over the fence and clamber down the embankments leading to the river. Over several trips, he hauled the logs down the steep, 30-foot embankment. The ground was soft and slightly unstable underfoot, and susceptible to wind. Wind was not necessarily a good thing for mushrooms, but as the perennials surrounding the logs grew taller, they might afford some measure of protection. More to the point, he had chosen this spot because it was relatively accessible, it was level ground, and it was wooded, providing shade to mushrooms that would not thrive if they dried out in the sun. The wild leeks, blue cohosh, and Solomon's seal, all perennials, were pushing up from the earth, as were ferns, still tightly coiled. The wild leeks crushed underfoot gave off a raw garlic smell. The river looked brown and silty, the consequence of spring runoff. These embankments had been the dumping ground for the farm's refuse, and it was not uncommon to find pieces of nineteenth- or early-twentieth-century crockery, disgorged by the earth as it shifted with each winter.

Kyle stacked the logs in a kind of cross-hatching pattern. The spawn should start growing immediately, although nothing would be apparent to the naked eye for a while longer. Closer scrutiny would then reveal the fuzzy, white mycelium, the indication that the mushrooms had taken root. He predicted the logs would soon be clotted with them, mushrooms springing up everywhere. When he brought down another batch of logs in another week—he had ordered some more spawn, but this time in the less expensive sawdust form—he would check on this stack of logs. As a further precaution against too much sun in the early spring, he would cover them with Remay, a synthetic, protective covering that admits air and sunlight but shields tender transplants from too much sun or frost.

Kyle's shiitake experiment seemed characteristic of the attention he gave to managing the resources of the farm as a whole, in particular the wood- and pasturelands. Over the course of the five years they had been on the farm, Kyle had inventoried the species of trees directly on and abutting their property, and had thought long and hard about how to best use them. Over time, he had devised a plan in which the trees were managed for sugaring and firewood and as sources for his woodworking. The best and strongest trees were allowed to grow to timber, while the weakest were allowed to die off, providing habitat for cavity-nesting birds, like wood ducks or woodpeckers, and when they finally fell, they were home to salamanders. Eventually they would disintegrate entirely, turning back into the soil. Whatever land wasn't used for crops or pasture was converted to habitat for bluebirds and tree swallows.

It was nearly the end of April. The fine, cool mist that had been drifting across the fields all morning had turned to a heavier, soaking rain. Good plant weather, and ideal for the chore that Jennifer had set herself that morning: transplanting thousands of onions from their seed trays into the ground. Rain was good—the farm needed the rain. There had not been much snow after December and January, and the previous year had been one of drought and parched earth. Let it rain this day and the next and the day after that. Looking out at the sodden brown earth and matted fields and bare trees, from which no hint of green emerged, it seemed impossible to Jennifer that in a matter of a week or two, the animals would be eating grass 6 to 8 inches high. And yet it happened that way, in that time, every year, and she marveled at it.

Onions, although not normally thought of as such, were fussy. They did better if they were planted from seed, not as the bulbs, or sets, that they were normally sold as. They didn't grow well

Kyle checks the progress of shiitake mushroom spores he planted in hardwood logs earlier in the summer. Because of a rain-soaked summer, the logs near the riverbank didn't need to be irrigated. Most mushrooms take a year to fully germinate, but one mushroom appeared ahead of schedule during the wet weather.

in the house, they had to be started in a greenhouse. Accustomed to the dim light of the greenhouse, and to the potting soil in which they had been started, they would recoil if you put them out immediately in too bright light and in soil that was alien to them. Jennifer had moved them out of the greenhouse the previous Saturday, to habituate them to the new environment. In a way, they were prone to the same complications as human organ transplants; they could easily reject their host, and it would be touch and go until they rooted and took hold. Onions, for mysterious reasons Jennifer couldn't quite put her finger on, sold in large quantities at the farmers' market. Mesclun, lettuce, Asian greens—the kinds of things that they expected to sell—didn't disappear as quickly as the onions did. So they had increased their onion stock in anticipation of selling out.

Jennifer would not have to transplant by herself, fortunately. Her sister Anna and her husband of two years, Rick Wilson, were up on one of their numerous annual visits. Anna typically made a circuit: her parents in East Middlebury, her sister Laura up near Burlington, her sister Jennifer in Royalton. Since 2002, Anna had been living with Rick in Milton, West Virginia, where she taught Spanish at a local grade school. Despite the fact that it was a seventeen-hour drive if you did it straight through, all the way from the West Virginia and Ohio border, she came up as often as she could because she was terribly homesick.

Anna was younger than Jennifer by two years. Like Jennifer she had long blonde hair, although hers was a lighter, almost platinum, shade that hung to her waist, while her sister's came to the shoulders. Unlike Jennifer, whose skin had been toughened and browned by so much sun and outdoor work, Anna's complexion was of a translucent pallor that would have been admired by a salon of nineteenth-century society ladies in Boston or New York. She was tall, taller than Jennifer, and wore glasses and had very pale blue eyes and pale lashes and a long face and looked like her mother, while Jennifer, with her squarish face and strong jaw, was the image of her father.

Both Jennifer and Anna wore slickers against the rain. Beneath Jennifer's slicker were garments that could only be described as "schmattas"—well-worn, almost threadbare pajama bottoms or sweatpants or sweatshirts with holes in them, the kind of clothes that not even the Salvation Army would touch but which were ideal for the kind of dirty work farming asked of her. No one would care if these clothes became encrusted with mud or manure; no one would even notice. Unless they were going to work off the farm, both Jennifer and Kyle dressed in this kind of haphazard, grab-it-where-you-can-find-it clothing, layers of old shirts and pants and sweaters and heavy wool socks and worn shoes of all description that bore absolutely no resemblance to the well-pressed, groomed, "country" look affected by well-scrubbed models in the Eddie Bauer or L.L. Bean catalog.

Brad picked his way carefully through the mud, down the slope from the barn. He was curious about a drowned chicken, looking from a distance like an unidentifiable white blob, that floated at the edge of the manure pit. Sometimes the chickens wended their way around to the back of the barn and down to the manure pit, which was the size of a small pond, and occasionally they ended up in it and drowned. Chickens couldn't swim and were prone to panic.

Despite such occasional accidents, the chickens were not penned inside or out unless they were pullets, chickens less than a year in age that had not yet laid eggs. (The pullets were kept separate from the laying hens which would pick on them because of their smaller size. Not for nothing is it called a pecking order.) In most cases, the chickens stayed close to the house and barn, didn't wander too far from safe harbor. They hid under bushes, they lurked behind the barn, they made nests on hay bales up in the haymow, they settled under the back and front steps to the house, and gathered in the lilies and phlox and iris and false indigo and hollyhocks and mock orange that Jennifer had planted on the barren slopes that led from the house down to the driveway, and which she was trying to turn into a garden.

Down in the upper pasture, Jennifer began to hoe long rows and, stooping over, moved down each row laying the tender shoots of onion in the shallow trenches she had dug. The onions were situated next to the garlic, which had been planted in the fall and mulched over with sheep's wool to keep down weeds and retain moisture. In the rain, the wool sent up the pungent smell of wet dog and mothballs. If you poked through the wool, you could see the beginnings of the shoots, or scapes, coming up through the earth. They looked yellowish, the result of burial under the protective

Jennifer transplants onions from the greenhouse into the ground. The farmers planted and sold out of 4,500 onions, which is 1,000 plants more than the year before.

covering. In a few weeks, the wool would be moved aside to expose the plants to more light and air. The previous year, the garlic had been planted down in the field closest to the road, where what looked like four crosses had been spaced at an equal distance from each other. It looked ominously like a small graveyard, but in fact these were trellises that held the raspberry cane.

Jennifer and Kyle rotated all the vegetable crops yearly, to rest the soil and to foil the assorted insects, which had something approaching institutional memory. For example, if you planted potatoes in the same spot two years running, you could be sure that while the insects may have taken some time to discover the plants the first year, they had no such trouble the second year. The solution for this problem is officially called "integrated pest management," and in the case of the ubiquitous Colorado potato beetle, apart from crop rotation, Jennifer and Kyle hand-picked the beetles and sprayed the plants with B.t. *(Bacillus thurengensis),* a fungus that attacks the larval stages but that is relatively non-toxic to animals and humans. They also did companion planting; basil and marigolds were planted near the tomatoes, for example, because the smell of the marigolds discourages some insects.

Brad was wandering around in the way three-year-olds have, wanting to be outside, intent on doing what he wanted to do, not necessarily what his mother or aunt wanted him to do. His attention wandered from the onions to the garlic to the rain to the dead chicken—a subject which he pestered Anna to explain. How exactly did the chicken get in there? Why did it get in there? Where was it going exactly?

"Brad, please go get the hoe," Anna said, trying to distract him from his interrogation, pointing to the slope that rose up to the barn, where he had heedlessly dumped the hoe on his way down.

"We need it to plant the onions," said Jennifer.

"It's bad for boys to get it," Brad asserted.

"A woman's supposed to get it?" Jennifer asked incredulously. She and Anna looked at one another, and frowned; *Hmmmph,* issued from both of them at the same moment. A broad grin crossed Brad's face.

"You're the worst of men," Jennifer said.

"You're the worst of women," he snapped back.

"You're the best of men," she said cheerfully.

"You're the best of women," he said, and brought her, by way of recompense, a tray of onions for transplanting.

"Oh, thank you, Baby," Jennifer said (pronouncing it *Bee-bee*).

All things considered, this was not a bad time of year. Things were, she felt, manageable. The animals were not yet out at pasture, although they would be put out in the next few weeks. Lambing was done with. There weren't any apprentices living on-site; Kathleen had left on April 10 to go work at a farm in Massachusetts. The daily schedule was not crushing. There was no weeding or plowing or haying to be done. They hadn't started going to farmers' markets. The CSAs didn't begin until the end of May. The planting was imminent, but that was nothing Jennifer minded. They were moving in slow motion, waiting for spring, which in Vermont does not arrive full-blown, as it does south of here, but unfolds gradually, even tortuously, leaf by leaf, blade by blade, flower by flower.

For all intents and purposes, they had made an alliance with nature, because there really wasn't any other way to do it—at least not in the way they had chosen to farm, without recourse to equipment and chemical fertilizers and pesticides. Nature imposed order on them, not the other way around. They could trick it, by using greenhouses, or try to outmaneuver it, by breeding animals out of season, or stay one step ahead, by planting vegetables in such a way that certain pests were discouraged, but they were essentially following and learning from nature's mandate.

Late autumn would be the beginning of dormancy; winter hibernation; and early spring an almost imperceptible awakening. Each May through September was an explosion of growth, a profusion of tasks and flat-out, nonstop hard work. Rather than say that Jennifer and Kyle lived in harmony with nature—a rather vague, insipid phrase—it would be more accurate to say that they took their cues from nature, that they'd learned the natural order worked quite efficiently if left alone, and the best one could do was to mimic it, or complement it. Which took considerable skill and knowledge: "In organic farming," Jennifer observed, "sometimes you just have to sit on your hands and think about what needs to be done."

By nightfall, the rain had stopped. Fog rose from the river and wafted across the roads. There were oceanic puddles on the dirt roads, which, not yet fully thawed, had been unable to absorb so much water. The night air was damp and chill. There was no moon because of the cloud cast, and it was as black as pitch. Jennifer, Kyle, Anna, and Brad were about to embark on something called Herp Night, an annual event undertaken in the spirit of exploration, in which they set out after sunset to see what might be moving about in the way of frogs, newts, or salamanders. (*Herp* was their shorthand for "herpetology," the branch of zoology that studies reptiles and amphibians.) The vernal pools that dotted the roadsides and woodlands—"roiling pools of salamander sex," Jennifer called them—should be full of all manner of amphibians and their eggs; creatures that spurned daylight and ventured out after dark, sometimes in great, pullulating tides of migration. Today's deluge portended herp weather; the frogs and toads liked to move in the rain. You could hear the spring peepers singing at full bore, calling to potential mates, and their music was shrill and jangling, of a pitch and timbre that penetrated the eardrum. The wood frogs, whose song was deeper and more throbbing, a bass vibration that filled up the woods and sounded as if it was issuing from the bowels of the earth, were also in full cry. The language of love, or sex, made up in urgency what it lacked in melodiousness.

Jennifer was driving the Volvo, Kyle in the front seat, Anna in the back with Brad strapped into his child's car seat. Rick had decided to forgo the rustic pleasures of Herp Night, and was sitting on the couch, reading. The Volvo, which Jennifer had bought used, was on its last legs—more than 200,000 miles on the odometer—and there had been discussion about what they were going to do. They couldn't afford a new car; they couldn't even really afford a used car. Could it be repaired to the point where it wouldn't fall apart on them again? Only a month earlier, they had been driving down Interstate 89, near Montpelier, and the entire wheel had come off—the tire, the rim, the hub, and part of the axle—sparks arising from the friction between highway and car. That they hadn't had an accident had been a minor miracle. The car never worked properly after that, even with repairs.

During the expedition, its headlights would fail twice, leaving them to drive down normally busy Route 14 without any lights at all, and the brakes would work only fitfully, after Jennifer and Kyle performed a kind of voodoo on the car to get it to go, which involved thumping the dashboard, hard, and alternately yelling at it, and pleading with it. Because it was after eight o'clock, and perhaps because of the weather, there was hardly any traffic. This was the country, after all—the sticks, the boondocks, people already tucked into their houses for the night. The car lurched along. They would make a loop, down Morse Road to Dearing Road, where there were some vernal pools just off the road, then down to Route 14 and back up to the farm. The car's headlights picked up what looked like small, glittering pieces of reflecting glass in the road: toads or frogs in mid-migration.

Whenever they spotted one, they stopped the car and Kyle or Jennifer got out to look. A large brown American toad sat inert in the middle of the road, almost directly in front of the car. Kyle picked it up and examined it, and then passed it to Jennifer, who passed it to Anna, and then to Brad. They restored it to the side of the road, and continued, stopping every few minutes to investigate possible sightings. At last, they pulled off the side of the road. Was there a vernal pool here or back farther? There was some tramping back and forth, flashlights waving about, and joshing about who remembered best where it was last year.

"Here it is," said Jennifer, pointing to a pool some ten feet off the road. They waded in. Kyle carried Brad on his arm. The ground was soft and muddy, squelching beneath their boots. They shone the flashlights at the pool, which was about 4 feet in diameter and clear, but clogged with fallen leaves from the previous autumn. On its surface was a floating grayish scum. Salamander eggs. Jennifer waded into the pool—the mud swallowed her boots and she would have to pull hard to get back out, as if she were in quicksand—and poked at the leaves with a stick, trying to dislodge whatever might be lurking there. There was a wriggle, a writhing something. She watched, waited, and then put her hand into the water and, pulling up a salamander, yelled "Spottie!"

After an overnight downpour, a pullet Cornish Rock hen negotiates a puddle. The wet summer benefited the crops, and because the farm doesn't rely upon mechanical equipment, wet fields didn't slow down any machinery.

The salamander that she clenched in her fist was about six inches long, gleaming black with yellow spots—quite beautiful in its own secretive way. She and Anna, and then Kyle and Brad, examined it with the close attention of high school science students before she put it back into the pool. It dove beneath the leaves. Farther up the road, at a crossroads, they walked down another road to a pond. Even when their eyes had grown accustomed to the dark, they had to rely on their flashlights, because the night was so dark, with not even a hint of light. This pond was full of peepers. You heard them before you saw them. Given the noise they emitted, they were astonishingly small, only a few inches long. They floated on the surface, near weeds and vegetation, where they were camouflaged. Kyle shone a flashlight on one so that Brad could see it; it took alarm and swam off, its back legs pulsating as it propelled away.

They went back to the car, but the sound of their voices and the sight of their flashlights had aroused the curiosity and suspicion of the elderly couple who lived in the tidy red clapboard house, more of a cottage, really, at the crossroads. You could see curtains being pulled aside, and faces peering out. Jennifer waved and called "Hello" cheerfully.

"I don't think they heard you," Kyle muttered. He hoped they wouldn't call the cops, as they were apparently wont to do whenever something even remotely out of the ordinary occurred on their watch. Everyone was back in the car before this could happen. Within minutes, Brad was fast asleep, breathing gently and evenly, a toy car clutched in his hand. On Route 14, heading south, headlights having ceased, temporarily, to function, they pulled over. On either side of the road, set back not too far, were two ponds. Jennifer rolled down the windows, and through them you could hear, at great volume, the calling of the peepers.

May 9 was pasture day, the first day in months that the livestock would be let out. There were nearly a hundred sheep, and eight cows and calves to put out to grass. Once the sheep pens were emptied, the sows, and their piglets, would be moved into them, and the pigpens, in which they'd been lolling and grunting over the winter, would be cleaned out. Kyle and Jennifer hoped one day to provide some kind of outdoor enclosure for the pigs, so that they could take some daily sun. They had learned, through bitter experience, what happens when pigs get out to pasture. When they had first owned them, the pigs had gotten out accidentally and had then run roughshod over the land—"Huge four hundred-pound blobs bucking through the fields," Jennifer remembers—gouging it up in their attempts to eat whatever hove into view. Since then, the pigs had always been confined in the barn.

First Jennifer and Kyle would bring out the cows, and then the sheep, and then they would deal with the pigs. Bill Fielding had started work. Jennifer untied the cows from their stanchions, one by one: Tildy Ann, Prism, Brown Sugar, Buttercup, Lucy (a calf that had been deposited there by Gene and Laura Craft), and the veal calves. They were deeply reluctant, even suspicious. "C'mon, c'mon, c'mon," Jennifer said impatiently. Kyle and Bill were on their back ends to push them if they refused to budge, which they did. Bill had the look on his face of someone who knew these were big, unpredictable animals and he was standing right behind them and, if they should so choose, they could kick out at him with utter impunity. How was he supposed to act? With authority? With respect? He took his cues from Kyle.

"Lucy, c'mon. Brown Sugar, you know what you're supposed to do," Jennifer coaxed.

They were led out past the pig and sheep pens, through a door into the Old Barn, and then through another door into the Horse Barn, which funneled out to the pasture. The door to the pasture was wide open. Freedom was theirs, but they didn't quite know what to do with it; for the last four and a half months they had become habituated to close quarters and constraint. At last, one cow, probably Tildy Ann, the de facto matriarch of the herd, bolted through the door. The rest stampeded after her, trotting clumsily and tentatively down the muddy slope to the first pasture. As the realization took hold that they were not penned in but had latitude of movement, they began to buck and kick through the grass in a kind of bovine jitterbug, more antic than graceful.

"I wonder if Tildy is lame," Jennifer said, watching her. "She seems to be limping." She revised her opinion. "I think they just haven't moved for four months. It's the most exercise they've had since the twenty-seventh of December."

Cody the border collie watches ewes and their lambs as they look at the spring grass for the first time. The eight-year-old dog has always been a pet for Kyle and Jennifer, not a working dog.

She turned to Kyle, conscious that his time this morning was limited. "Honey, when do you have to leave?"

"Soon. Let's go," he answered. He was on the schedule to drive the school bus.

They returned to the barn, where bedlam reigned. Now that the cows had gone, the other animals knew that something was afoot. There was a kind of mob mentality at work, a mass hysteria. The ducklings, which were kept in a large crate in the aisle between the sheep pens, were quacking frantically. The sheep were in a panic, all baaah-ing at once, uncertainty written on their faces, their jaws opened to release a pure stream of sound. The pigs couldn't have cared less. Kyle climbed into the sheep pen nearest the door that led into the Old Barn. Bill was on the gate. The sheep, seeing Kyle in their midst, piled on top of each other in the farthest corner.

Kyle got behind them and clapped his hands, to drive them out. The lambs, which had never left the confines of the pens, and had not been separated from their mothers since they were born a few months before, were in a frenzy of anxiety. They had no idea what to do, or how to behave. Bill and Kyle succeeded at last in funneling them out of the pen and through the door, but one lamb remained in the pen, running from one end to another as Kyle tried to corral it. Finally, in a weird synchronicity, the lamb literally leaped up into his arms, just as Kyle tried to grab it, as if Kyle were his savior and not his captor. The sheep streamed through the Old Barn, into the Horse Barn, and out the door.

Once outside, however, the sheep persisted in huddling as if under threat. It took the cows only a few minutes to understand that they could move; the sheep hadn't yet grasped this new and essential fact of their lives. They baaah-ed so loudly and persistently that it sounded like a kazoo orchestra, razzing and loud and dissonant. A small group of lambs clung, limpet-like, to a corner of the barn. Kyle, slightly red in the face from the exertion, scrutinized them, and then their mothers, which had left them behind and were following Jennifer, not realizing that their lambs were far behind them.

"This is when the ewes'll blow through a fence," he said. "When they're looking for their lambs and don't know how to get to them."

Jennifer had been leading the sheep to pasture, but when she saw that the lambs were refusing to move, she led the ewes back up the slope, past the planted onions and garlic, past the manure pit, toward their young. "C'mon girls, c'mon girls, c'mon girls," she hollered. They trotted behind her. Some of the lambs, seeing them, began to leave the huddle and follow their mothers, but success was only temporary. A renegade group of lambs broke away from the ewes and scrambled back up the slope to the barn, where they gathered stubbornly in the same spot.

"It's like herding worms or something," Kyle said, watching the lambs wriggling this way and that.

Bill tried to herd the lambs down and away from the barn, but there were too many of them, and their flocking instinct too deeply ingrained. They functioned as a group, not as individuals. They would not move, unless somebody—human—or some animal—ewe—showed them it could be done, and even then it was a leap of faith for them, one that they made timidly. Kyle was looking at his watch. He had about fifteen minutes before he had to get ready to leave, and the sheep were in chaos, scattered every which way.

Jennifer went up to the barn and returned with a bribe: the all-important grain bucket. She thrust it in front of the lambs, which sniffed at it. Moving slowly, and clucking encouragement, she waved the bucket and they began to follow, as if she were the Pied Piper of Hamlin. When they saw their mothers, they broke into a gallop. Jennifer led them through an opening in the electric fence, and out onto pasture for the first time in their lives. The day was overcast, but warm, and the grass was inviting. When you see lambs at pasture, you understand what is meant by the old-fashioned phrase "to frolic and gambol," for that is exactly what they do. They fling themselves about, they spring and bounce on their long legs as if they were gazelles, they kick up their heels, they cavort. Some of them would be slaughtered for meat, others would be selected for breeding, and rarely, one of them might metamorphose from livestock to pet.

Ten days after they were put out to pasture, eight lambs, less than two months in age, and one goat were loaded into the back of Kyle's pickup. They were being taken to a slaughterhouse in Rutland. Five of the lambs would be sold as meat to Vermont Quality Meats, and three would be

Jennifer pushes a lamb out of the barn for the first time. The lamb's crying was only temporary—the ewe was put out to pasture next.

sold at the Norwich Farmers' Market. Because these lambs would be sold at retail, they had to be slaughtered in a federally inspected facility.

The lambs couldn't know what was going to happen to them, but they were made uneasy and unhappy by the separation from their mothers and from the farm; they baah-ed and bleated and shifted about nervously in the truck. It was a 45-mile drive to the south and west, up and over the Green Mountains and over a pass at Killington, about 3,000 feet above sea level, and down into Rutland, which is hemmed in by the Green Mountains to the east and the Taconics to the west in New York State.

Fresh Farms Beef was at the outskirts of town, off Route 7, down a paved road that turned to dirt. It was a squat white building, with an attached holding pen for the livestock. Workers, dressed in the regulation coveralls and shower cap, came outside on their breaks to smoke. When Kyle pulled up, backing up the truck to a holding pen, Jeff Nichols, one of the owners, a shortish, dark-haired man with big, Tweetie-bird eyes and a look of perpetual worry on his face, came out to help him unload the animals. Kyle got into the back of the truck to push the lambs out. Nichols looked at the goat, which was quite scrawny.

"That goat's pretty small," Nichols observed. "It'll cost you more to process it."

Kyle shook his head, and waved his hand dismissively. Kill it, the gesture said.

Fresh Farms had been at this site since 1996. It was a family business. Since moving to Rutland from their previous operation in Barre, Vermont, the slaughtering had been *halal*, performed in accordance with Muslim religious and dietary law. Both Nichols and his nephew, Nick Greeno, who also worked there, had converted to Islam, and a Muslim veterinarian would sometimes come in to ensure that animals were being slaughtered according to the laws laid down in the Qur'an, which are these: the animal must be conscious at the time of slaughter; it must be killed by cutting its throat; and the person doing the killing must offer a prayer prior to slaughter, "In the name of God, God is most great." Only then can it be said to be *halal*, or lawful. Using the halal method had allowed Fresh Farms to secure prison contracts in the Northeast, providing halal meat to prisoners who were Muslim or had converted to Islam while in prison.

Nichols's father had begun in the 1940s by picking up dead cows and selling them for dog food. Nichols, who was born in 1958, learned how to dress animals at an early age: "They gave us a steak knife at three or four years old, to learn how to carve." He was playing at dissecting a toy horse with a plastic knife when he was in his crib.

"This place was built in the nineteen-twenties, as a railroad spur," Nichols explained later. "They railroaded cattle from the West in the old days; the kids down from school would drive animals here for slaughter." For a slaughterhouse built decades ago, it was capacious and sturdy and well laid out. They handled hundreds of animals weekly, depending on the season. Federal law, which stipulated that animals must be unconscious at the time of slaughter, granted religious exemptions to slaughterhouses that killed according to Islamic or Judaic dietary law—halal and kosher—both of which require that the animal be conscious at the time of slaughter. (Earlier in his career, Nichols had worked in a kosher slaughterhouse in Montreal.)

A copy of the Qur'an sat on a desk in the small office where Nichols and Greeno and Karen Sears, the USDA inspector who was on-site weekly to certify the meat, entered data into a computer. The front room was papered over with posters and flyers and regulations. A placard for Biancardis, one of their clients down in the Bronx, read YOUR BUTCHER IS STILL YOUR BEST FRIEND. There was a copy of Zagat's 2000 and 2001 restaurant guide to Connecticut and southern New York State. There were photocopies thumbtacked to the wall: a post–September 11 American eagle; a sign reading, THE TERRORISTS HAVE WON THE TOSS AND ELECTED TO RECEIVE; George W. Bush talking to Saddam Hussein on the phone, both men holding receivers to their ears: "Can you hear me now?" Bush asks.

The slaughterhouse was divided into the kill floor, the freezers, and a processing room, where workers dressed the meat, using an assortment of lethally sharp knives to carve the meat, whether beef or veal or lamb or goat. (They did not process pig, which is considered unclean in Muslim

Michael the guard horse keeps an eye on his flock of sheep. A retired national champion riding horse, the Anglo-Arabian keeps predators away from the sheep.

dietary law.) Nick Greeno was cutting up a cow hanging from a hook; he worked with practiced expertise, slicing rapidly through the cow's rib section. Piece by piece, the carcass was dissected, as if you were taking apart a house, section by section. He wore, as did the other meat processors, protective covering from head to toe, and a shower cap.

The Fat Rooster lambs would be killed almost immediately. From the moment an animal was brought from the holding pen and through a door onto the kill floor, it had, in effect, ceased to be animal and was now meat. There is a kind of art to killing an animal. It is, as Jennifer later observed, an oddly intimate act, one requiring great skill. Once you have embarked on the act, you cannot flinch from it. Hesitation and deliberation and hand-wringing have no place, because that only prolongs the animal's suffering. Sentimentality may make you feel better, but moral equivocation means nothing to the animal; unless, of course, you choose to spare its life, an unlikely eventuality when you are speaking of livestock.

The kill floor was 20 by 40 feet; there was a hook and pulley system hanging from the ceiling, from which they hung the dead animals. Once the lambs were slaughtered, they were hung upside down from hooks to drain the blood, a process called "bleeding out" that decreases both the growth of microorganisms within the blood and the risk of spoilage. From there, they were moved to a cutting table, where they were skinned, their coats thrown onto the floor and reserved for the wool. A series of lambs, skinned, hung on hooks not far from the freezer; Nichols slit open the first lamb's belly and took out a bag containing the lymph nodes, the liver, the heart, and the lungs, which plopped to the floor in a pudding-like mass.

Karen Sears, the USDA inspector, stood between the kill floor and the freezer. When an animal had been killed, skinned, and gutted, she looked it over for any abnormality, "things that are plain to your eye," she said, like rabies or foot-and-mouth disease. "Any disease should show up in the lymph nodes. A lot of times livers are bad because of bad water." If she determined that there was a potential hazard, a veterinarian would be summoned to make a diagnosis. But, she added, "Very seldom do you see problems in lambs less than forty days old."

"You carry 'em, I'll cut 'em," Nichols said to a Latino worker, Faustino, who wore a cross around his neck. Both men wore bright orange coveralls. Faustino vanished through the door and carried in the first Fat Rooster lamb, his arms around its belly as if he were a child hugging an oversized rag doll. The volume at Fresh Farms was not large enough for an assembly line, and bringing them in singly reduced the animal's stress, which butchers have known for hundreds of years to be a key component in the quality of the meat. (In *On Food and Cooking,* food writer and scientist Harold McGee notes that the more agitated the animal, the more likely, because of chemical reactions within the body at and post-slaughter, that its meat will be tougher, gummier, and grayer-looking, a phenomenon known as "dark-cutting.") Here, the sheep were not exposed to the agitation of other animals and they were not watching other animals being slaughtered, the sight of which they might not comprehend, but the smell of which—blood in quantity has a metallic, acrid odor that overpowers almost everything else around it—they certainly would.

The first Fat Rooster lamb made no struggle or protest, no noise of any kind, and showed no fear. Faustino held it still. Nichols approached it and, with one decisive thrusting motion, stabbed the knife deep into its throat, slashing swiftly from its right to its left; Faustino averted his face at the moment of killing, to avoid being sprayed with a quantity of blood. As it was, there was a streak of blood on his cheek. The time elapsed was a minute or less—very fast. The lamb was transferred to a hook to hang upside down, where it went into the reflexive neural and muscle convulsions that follow slaughter. There was a gagging, choking sound, the sound of air rushing out of the lamb's windpipe. Until an animal is all but drained of its blood, which carries the oxygen to its organs and nerves, its system continues to function in this automatic way, and even when it has bled out, it is not unusual to see muscles twitching involuntarily.

After each slaughter, the floors and the carcasses themselves were hosed down; the less blood, the less chance for bacteria.

"I had a dream last night," Karen Sears told Nichols. "I moved back to Texas. On a clear day in Texas, I could see for ten miles." Sears had been an inspector at Fresh Farms for six years; although originally from Arkansas, she had worked in Texas as an inspector at a cattle processor, where, she said, "I saw more animals on one shift than I've seen in six years here: twenty-five hundred cattle in one shift." She turned the lamb carcass from side to side, examining it. When she was done, Faustino stamped it with a USDA seal, and marked it. Sears looked at the carcass critically. "You're supposed to be able to read 'em, Faustino," she said, pointing to the USDA seal, which appeared a bit blurred. And so the slaughtering continued through the morning, one lamb after another, each one quickly dispatched, skinned, and eviscerated, their wool reserved for later use.

Both Jennifer and Kyle had reached an accommodation of sorts with slaughtering. Jennifer had learned how to slaughter livestock when she was living in Maine with Brad and Donna Kausen; she and Kyle were raising lamb and chicken to serve at their own wedding reception. She began by slaughtering lambs. "We did it with a gun, and I did not like that at all. Then we did it with a knife. I don't know why, but it was not as bad.

"It's hard to look at it," she reflects, "having the knowledge that you have the power to kill something; it's not actually *doing* it but that I *can* do it. It's not hard to slit its throat, but it's hard to think that I *am* slitting its throat. Kyle didn't like it. It took him two or three years before he could do it. If you don't do it the right way, if you don't get the jugular veins, they don't die as fast.

"It feels like ripping a sheet. That's the kind of force you have to have to do it. You know when you're deboning a chicken, and you're making a cut between the joints? That's what it feels like."

"We didn't slaughter at home when I was a kid," Kyle recalls. "It was all done commercially; we shipped cattle out by the truckload. I did visit other neighbors when they were slaughtering, but I didn't do it until we went to Maine and helped slaughter chickens and lambs. Sometimes it was all right and other times I just didn't want to do it. Brad and Donna helped teach me, and I also read books. It's one of those things where reading a book about it doesn't help that much. Intellectually, because we raise and eat meat, it would be completely hypocritical to refuse to participate. It's awfully convenient to send it somewhere else. You feel for the animal when you're slaughtering it, but not enough to not slaughter it. Some people talk about the power they feel. I don't feel that. That's for somebody else."

On the Thursday before Memorial Day weekend, Jennifer and Kyle and Bill Fielding, as well as Gene Craft, were over at Luna Bleu farm in South Royalton, which has the look of many farms in Vermont: slightly down-at-the-heels but functioning. There were twenty-five ducks in the back of Kyle's pickup truck, which would be killed this morning. Jennifer and Kyle wanted to sell them at the Norwich Farmers' Market on Saturday, May 24, which would be the first big market of the season; people had already been asking for duck.

Brad was roaming around the barnyard. Kyle looked over at Jennifer. "He shouldn't be here," he said sternly, knowing that Brad tended to become upset when he saw animals or birds that he considered to be pets, killed. Jennifer protested that there was nowhere else to take him. As it was, Brad wandered around not paying any attention to the slaughtering at all; he was far more taken with a dog that lived on the farm, and followed it around, and explored the barn.

They had set up, next to the barn, the killing cones—the inverted cones into which the bird is put, head and neck first. The cones were tacked to a side of the barn; a plastic sheet hung below the cones, which they would hose down once the slaughtering was done. Once the ducks were dead, they went into a scalder, which, with the addition of dish soap in the water, softened the feathers. From the scalder they were immersed in a plucker, also filled with warm water, which theoretically removed the feathers—although duck feathers, unlike chicken feathers, are notoriously stubborn and difficult to pull out. There were two plastic tables on which they would eviscerate the ducks; one bucket, in which they would discard the guts; and two pails filled with cold water into which they would plunge the plucked, eviscerated ducks. Following slaughter and dressing, all animals or poultry must be kept cold, below 40 degrees, to prevent decomposition and the spread of bacteria.

Jennifer carries a duck to be slaughtered at Luna Bleu Farm in Royalton.

The ducks were ten weeks old, two weeks older than they should have been for slaughter, and some of them had begun to grow adult feathers, which made them harder to pluck. Kyle would cut the ducks' throats; Jennifer, Bill, and Gene would pluck and eviscerate. Kyle reached into the back of the truck with a croquet mallet to separate the ducks from each other; he took two at a time. Every time he did this, the ducks sent up a loud, cacophonous quacking. "They make a happy duck sound in the barn when they know they're getting food," Jennifer said. "Now they're doing the alarm sound. They know a guy isn't reaching in with a croquet mallet to feed them."

As soon as Kyle removed a pair of ducks, the quacking in the truck subsided—temporarily. He carried the ducks over to the killing cones, thrust them headfirst into the cone, and drew their necks down through the opening. They struggled. He first pithed them, driving a sharp knife into the brain, and then with a small sharp knife he cut across the jugular vein; as he did so, one duck emitted a last faint strangled peep of pain and surprise.

"The pithing supposedly causes brain death," Kyle said. "It removes the sensation of pain. It's supposed to stop movement of the muscles, stop the contraction around the skin. That's what people say causes less pain; but then," he paused, "I'm not a duck." The ducks, throats cut, thrashed in postmortem convulsions, necks arching, wings thrumming against the cones. Then they fell still, their necks straightened, and they hung limp, blood dripping from the neck. It might take only three to four minutes for them to die, but it seemed interminable. Life is not extinguished, or relinquished, so easily.

Gene Craft, who had grown up in rural Virginia, where his grandfather had a farm, recalled visiting on the day that his grandfather cut off a rooster's head and it ran around the yard. "There's nothing worse than the feeling you're not killing the duck or chicken," he said in a voice that retained a hint of a soft Southern lilt. "Some people chop the heads off, severing the central nervous system. They move around more, but I don't think they die any faster than having their throats cut. It's harder to get geared up for killing than doing it."

Indeed, this is true. What seems at first difficult to watch becomes matter-of-fact; a gradual desensitization takes place. If you farm, and you have livestock or poultry, this is what you do. If you eat meat or poultry, this is how it is obtained.

Kyle methodically removed the ducks, a pair at a time, until only four remained. "He's made it," Jennifer said, "so there isn't one duck left standing, but two."

"For any herding or flocking animal, that has to be the biggest fear," Gene observed. "To be singled out, to be left alone."

The last four ducks in the pickup truck were scattering, trying to get away from Kyle. Jennifer made a last-ditch plea to spare the lives of the last two ducks. She had already spared two back at the farm: one because it had what can only be called a pompadour, an abnormality in which a patch of its head feathers fluffed up instead of lying flat. Jennifer couldn't bring herself, she said, to kill a duck that wore what looked like a bonnet. The other was simply too small to kill, a runt.

"We could put these ducks on the pond," she said to Kyle, although, in truth, it didn't seem like a serious attempt to spare the ducks, but more of an afterthought, and perhaps, subconsciously, a deliberate irritant to Kyle.

Kyle looked at her, his face taut. "You know what I think, and you know what I'm going to do. Don't make me out to be the villain. I wouldn't have had them in the first place."

He killed the last two ducks.

"Now I'm going to have nightmares," Jennifer joked.

"And mine will stop," Kyle said.

THE LAND

June had been a month of little rain. There was worried speculation that this summer was going to be like the last summer and the one before that—drought, weeks of it, and unrelentingly hot weather. Corn that was stunted, with burnt tassels. Cucumbers that wilted under the glare of the sun. Greens that withered away to nothing. From a farmer's standpoint, the Northeast poses its own particular challenges—rocky soil, cold climate, short growing season, a dearth of available and affordable

Kyle picks beets with apprentice Meg Wedding. Meg spent sixteen weeks at the farm—she and her boyfriend, Ben Canonica, found the farm through the Vermont Fresh Network. Their goal is to someday work a farm of their own.

arable land—but unlike the West—west of the 100th meridian, where rain is measured by the table-spoon—it has always been blessed with plentiful rainfall and an abundance of potable water.

Prolonged drought is not something to which farmers here have been accustomed, and even if they've gone without rain, the water table has been fairly resilient, enough to sustain them during a dry spell. Not so these past few years, when water levels everywhere had dropped to alarming lows: rivers drained almost to riverbed, ponds whose waters had shrunk down to fetid mud bottom, wells that no longer delivered water. Even the waters of Lake Champlain had begun to recede.

Some farmers irrigated their crops; Fat Rooster never had, depending entirely on rainfall. Because they had drilled a well, they now had the capacity to irrigate, but the expense of irrigation is prohibitive, and the labor involved more than Kyle and Jennifer wanted to do at this point. Nonetheless, they had tripled production this year, on the assumption that they would have the apprentices to help and, they hoped, sufficient rain to bring the crops along.

April and May were rainy—sodden, actually—which was a good sign. The first significant Norwich Farmers' Market of the year, held the Saturday of Memorial Day weekend, was a soaking downpour, which depressed, but did not drive away entirely, the loyal customer base. The Fat Rooster booth was on the south side of the market and was horseshoe shaped; you had to walk into it rather than strolling alongside it. This, Jennifer and Kyle felt, put them at a slight disadvantage. They had been coming to this farmers' market since 2001, and they had taken note of the fact that booths that are open tend to attract more customers than do the ones where you have to penetrate the space and engage with or confront the producer. They were still fairly new to the game. Other farms, like Luna Bleu, had large followings; Fat Rooster was just beginning to draw a core group of customers.

The market opened in early May and ran through October. The first few markets were notable for sparser attendance and less produce for sale. It was too early to show a cornucopia of vegetables and greens; they were not yet ready for harvest. This Memorial Day weekend, Jennifer had set up on the booth's shelves, tins of maple syrup, bunches of wild leeks, cartons of fresh eggs, some flowers and seedlings for sale in trays, and in the meat freezer, lamb and duck and veal, which were snapped

While Meg enjoys a lemonade from a fellow vendor at the farmers' market, Jennifer tells a customer how to prepare one of their produce items for sale.

up quickly by customers. Fat Rooster Farm was one of a few farm stands this year to offer organically raised meat; there was a man who sold domestically raised elk meat, and he stood a lonely vigil. The rain had let up, was now more of a drizzle, but saturating nonetheless. The pickup truck was parked behind the booth, its back open so that they could move the wares directly onto the shelves and tables in the booth.

The Norwich Farmers' Market is on a half-acre plot of land off Vermont Route 5 in Norwich, a wealthy town of nearly 3,600 people directly across the Connecticut River from Hanover, New Hampshire. The parking lot at the market looks like an ad man's fondest dream, a cliché really, overflowing with Volvos and Range Rovers and Audis and Subaru Outbacks and Saabs and Jeep Grand Cherokees; the battered pickup trucks belong to the farmers. There are some 100 active vendors, selling food or produce or organically raised wool or baked goods or pottery or flowers. There is a small gazebo at its center in which musicians play—what else?—chamber music or folk music.

Five miles to the south of the market, at the confluence of the White and Connecticut rivers, is the raffish railroad town of White River Junction, a town that, for the last twenty years, has been threatening to have a renaissance and never quite achieves it; the forces of gentrification are in a war of attrition with the deeply rooted, ineradicable forces of poverty and malaise. This stretch of Route 5, which parallels Interstate 91 (both north–south roads that run the length of the state), used to be farmland and is now quasi-suburban, with subdivisions with names like Tomahawk Village and Hemlock Ridge; Norwich Commerce Park; and a mix of businesses including a Subaru dealership, the King Arthur Flour mill, three dental practices, and Killdeer Farm Stand.

In money and in education and in their expectation of what life holds and what is due them, the upper-middle-class consumers whom this market attracts are as far removed as it is possible to be from the dilapidated apartment buildings on South Main Street in White River Junction, or the dairy farms twenty miles north on Route 5 in Fairlee and Bradford, or the trailers and modular homes up in Corinth and Tunbridge and Royalton, where the TAKE BACK VERMONT signs—Take Back Vermont from civil unions, from taxes, from school funding; wrest it away from the flatlanders and the gays and the tree-huggers and the liberals—are still prevalent, even three years after the bruising political fight for governor between the Democrat, Howard Dean, and Ruth Dwyer, a Republican, that brought them into being.

The Norwich Farmers' Market is, figuratively, the northern extension of Manhattan's Union Square Market, not Farmer Hodge's homespun Farm Stand and Christmas Shop up in Fairlee. Yet it serves an important purpose: It offers a steady, profitable venue for farmers, and it brings consumers into direct contact with growers. Bonds are formed and an informal education is offered on how food is grown, and perhaps the people who don't farm will see the larger value of the people who do. In 2003, Fat Rooster Farm would pull in from this market alone more than $10,000 in gross profit, which made it well worth the effort and the travel once a week.

Ben Canonica and Meg Wedding were behind the booth, wearing slickers. They had arrived at the farm a few days earlier. Both were blond and slender, with honeyed coloring and the freshness of youth and the same friendly, unassuming way with people. There was a similarity there; perhaps that was what had drawn them to one another. Ben and Meg were still in the first flush of romantic feeling and affection, although they had been together for nearly two years.

Ben hailed from Boxford, Massachusetts, in the northeast corner of the state, where his family had what he called a subsistence, or hobby, farm, not the kind of farm from which you try to earn a serious living. He was twenty-five, Italian-American, tall but lean, with the blond, finely drawn, aquiline look that many Northern Italians have, already losing his hair, which emphasized the high dome of his forehead. Meg was twenty-four, with the looks of an ingenue—button blue eyes, white teeth and freckles, brown skin and rounded limbs. The kind of spirited, clever farm girl who, in a Willa Cather novel, would be the one that all the other boys were sweet on but could never quite catch up to because she was just that much quicker than they were.

She grew up in Centerville, Indiana, ten minutes from the Ohio border and an hour-and-a-half drive from Indianapolis; her grandfather and great-uncle were farmers. She met Ben at a ranch out in Colorado in the days immediately following the attacks of September 11. She had been working

Jennifer walks to the barn from the garden with Cody. She is carrying squash blossoms requested by a customer at the Saturday farmers' market.

there that summer and was on her way home, but the grounding of the American air fleet, from September 11 until 13, delayed the return trip. Ben had worked there the previous year and was coming just as Meg was supposed to be going.

Their aim was to have their own farm one day, which was dependent on whether they could raise the capital. Their apprenticeships had served the cautionary purpose of both dissuading them from the burdens imposed by farming, and persuading them of its value and necessity. "We both want to learn as much as we can," says Meg, "but with land prices it seems as if we need to stop working and earn money in order to do it." Between them, they had worked on enough farms, says Ben, to "figure out what you want to do and what you don't want to do." They were both leaning toward staying in New England, Ben because he grew up here and knew it like the back of his hand, and Meg because it seemed, after all the places she's worked, the most agreeable to her.

Meg liked Colorado but wasn't crazy about the fact that everything had to be irrigated: "It's so dry. To me, sustainable agriculture isn't really sustainable when you have to irrigate everything." She went to college in North Carolina, where she studied agriculture, and liked the landscape but not the summer heat. The Fat Rooster model seemed the closest to the kind of farm they would like to operate, in terms of the size and the range of things done.

Ben and Meg had only been at the farm a few weeks by the time Memorial Day rolled around, and Jennifer was smitten. "They could run the farm without us," she said without hesitation. This initial impression was only borne out by subsequent interaction. Fat Rooster was many things, but punctiliously organized was not always one of them. Ben and Meg, however, *were* punctiliously organized. They helped streamline the operations and they worked without complaint and they worked without ceasing. Meg would plant or weed for hours at a time, as would Ben. They were almost always cheerful and cooperative, which impressed Jennifer and Kyle. They did many small favors but asked for nothing in return. The only thing that Ben wanted was the opportunity to work with Bobby, the draft horse. He wanted to learn how to use draft horses in farming, and since Kyle's first abortive attempts to use them to plow, Bobby, Ruby's foal, had never been put to the use for which he was intended.

On June 24, one of the first really hot days of the summer—92 degrees in the sun, and 89 in the shade—Ben was out on the tedder in a field a quarter-mile down the road, land leased from a neighbor, its sole purpose to provide hay, not to pasture animals. Once the hay had been tedded, it would be baled and stored in the haymow in the barn, and used to feed the sheep and cows through the fall and winter months.

It was one o'clock, well past the lunch hour, which at the farm was somewhere between eleven and noon, given the early hour at which they rose. As they had last year, Kyle and Jennifer had borrowed a tractor, tedder, and baler from a friend. The way it worked was like this: first, they mowed the fields; then they used a tedder, which was attached to the back of the tractor, to spread out the newly cut hay so that it dried evenly in the sun; and finally, they raked and baled it. Kyle did the mowing, Ben would do the tedding, Kyle and Gene Craft would take turns raking and baling, and everyone would help to put the bales in the barn.

This was the first cut of hay, early in the summer, and it was particularly fragrant. When the hay grew back, they should be able to get in a second cut in July or August, weather and equipment willing. Farmers called the second cut the Rowan, and to animals, Jennifer said, it was the equivalent of eating ice cream. The previous summer they had stored in their barn 2,300 bales of hay, all from their own fields and the fields they leased.

This year they could not have asked for better hay. The rains had pushed up the long grass in profusion, producing healthy, fat stalks that the sheep and cows would devour with relish. As it was, Jennifer and Kyle had a hard time containing them to their designated pastures; the animals had barely finished grazing one area when they broke through fence looking for the next lush patch of grass.

Ben was driving an older model of tractor with a tedder attached. Newer models of mowers, called mower-conditioners, are more efficient, simultaneously compressing and extracting moisture from the grass. With the older models, it was best to wait for the grass to dry, which was why Ben hadn't started mowing until after ten A.M., by which time the mists had lifted and the sun had warmed the

Lisa Koch, left, and apprentice Solange de Hautefeuille work to stack hay bales in the barn loft. Solange worked on the farm for six weeks, one of three apprenticeships as part of her studies at Istom, an agricultural school in France. Lisa and her husband worked in exchange for part of their Community Supported Agriculture (CSA) share.

earth and the morning dew had evaporated from the field. This was early summer heat, dry and tonic, not the sweltering humidity that cast a pall over the valley for at least a week and sometimes longer, in July or August, and sent people to the rivers and the lakes and swimming holes to cool off.

Ben drove the tractor, covered with a canopy to keep off the sun, in narrowing rectangles, from the outward perimeter inward. This field, which sat directly opposite Ella Hyde's farm, was longer than it was wide, a shallow depression in the earth that seemed to amplify the sun's heat. Rivulets of sweat trickled down his face and chest. The tedder, a rackety old thing, clattered by; it had four wheels with attached spokes that rotated at a furious rate, and stirred up the mown hay. It looked like the wake of a boat, churning up green foam. As it passed, moths flitted up out of the grass, disturbed by the clamor of the machinery. In its wake the tedder leaves scattered heaps of the long grass which, under the heat of the sun, smelled like honey.

At the edges of the field, in the shade, were goldenrod, burdock, clover, and milkweed. There was a rise near the river, and the tractor disappeared beneath it. Warm, dry gusts of wind tossed the mown hay in the air. After the field was mowed and the hay dried, Kyle and Ben would have to go through again and rake the hay into tidier windrows for baling.

Jennifer and Meg were in the barn mucking out the sheep pens. Brad was wandering around in the aisles of the barn. Kyle and Jennifer had hung a swing from the rafters for him, and he had his own tricycle and some toy trucks that lay in the aisles between the cow stalls and the pigpens and sheep pens. Another apprentice was working with them, a twenty-year-old French woman named Solange de Hautefeuille, from Dijon in Burgundy, where her family had a dairy farm. As part of her agricultural training at college near Paris, she was required to learn English, and so she had sought out a farm apprenticeship in the United States. Solange would be there for six weeks, until the end of July. She was not fluent in English and, perhaps because of it, was quite reserved, would say a phrase or two in uncertain English and then look away. She and Meg did not look dissimilar—blonde hair, blue eyes, snub noses—although Solange wore her hair in a bob, while Meg had hair below her shoulders that she usually wore up in a ponytail or a twist.

Jennifer had brought a portable radio into the barn; the lonesome, tinny quaver of Neil Young was singing, "Hey, hey, my, my. Rock and roll will never die." In the heat, the smell from the stalls was overpowering—waves of ammonia, even with the doors and windows open. The pigs had turned the hay in the pens into a tamped-down, teeming breeding ground for bacteria. The hay, now mixed through with pig manure, was as dense and solid as earth, its odor equally dense and solid. "The most important job is getting it out of here," said Jennifer. "It's like white ammonia smoke if you don't move it, because you're exposing the anaerobic bacteria. If we had all five of us, we could get this done in a day." By which she meant she and Kyle, Ben and Meg, and Solange.

Through prior arrangement, Bill Fielding had moved over to work at Luna Bleu, a move which Jennifer viewed with some relief. She liked Bill—he was genial and polite and good-humored—but he was not as focused on the job as she would have liked. He tended to wake up late, for one thing. "You don't wake up at eight o'clock when you're an apprentice on a farm," Bill would admit sheepishly. He was always contrite, almost to the point of self-abasement. She couldn't get mad at him because he always undercut her irritation by being so repentant that it would have seemed churlish not to accept his apologies. So he stayed out late with his buddies and couldn't get up at six or wandered away from the booth at market because he saw friends, but was then so hangdog, tail between his legs, that all she could do was shake her head and needle him lightly. He would probably be happier at Luna Bleu, which turned out to be the case; more people his age, and he wasn't asked to get up quite as early.

Solange had the tirelessness of youth, but the concentration of maturity. Like Ben and Meg, she would work from dawn to dusk without a murmur of complaint. Here she stabbed a pitchfork into the piled muck, wedging it in and breaking up the muck, as did Meg. All three women—Jennifer, Meg, and Solange—wore tank tops and shorts and heavy boots. The piglets, which were not really piglets anymore but young pigs, had been moved into the pigpens.

The three sows were all bred, due to farrow at the end of summer or in early September. "That pig was almost dead this morning," Jennifer said, pointing to a piglet. "They were walking on her. She

Ben uses a knife to cut mesclun salad mix. When cut, the plant can be harvested three times during the season.

wasn't eating or drinking. I stuck her with an eighteen-gauge needle and she didn't budge. If a pig doesn't get up for food, something's wrong." Jennifer had injected her with antibiotic and now she appeared on the mend. The piglet was on her feet, moving gingerly about, and clearly thirsty, as she went to stand under the tap in the pen, from which Kyle and Jennifer drew the water for the animals.

The radio was tuned to an oldies rock station, the kind that played music from the '70s and '80s and was manned by disc jockeys who, regardless of gender, all talked at the same relentlessly loud and frenetic pitch, as if they were parodies of themselves. A hoarse male voice, accompanied by a ragged guitar sound, erupted from the radio: Bryan Adams's "Summer of '69," the kind of rock anthem that is the staple of album-oriented rock stations; the sound of summer.

"Isn't this Bryan Adams's first record?" Jennifer asked.

"Yeah, this is where he should have stopped," Meg said, over the radio. In the background, Adams sang that there's no use in complaining when you've got a job to do.

"Then he sucked," Jennifer yelled. Brad sidled up to his mother—he had a way of appearing and disappearing—and pointed to the sheep pens, which, slowly, were being emptied of their refuse.

"Put the hay here, Mama," he demanded.

"We are going to put the hay here soon," she nodded.

The next day was as warm. This was the farm at its most fruitful; generous, expansive summer redeeming months of unforgiving cold and pinched ambition. You could smell it, a heady nectar of cut grass and manure and animal and subtler floral notes. The crops—row upon row upon row—were strong, healthy, green. The flowers Jennifer had planted were growing up as dense as jungle, splotches of red and blue and pink.

The light was brilliant, already high by eight in the morning and lingering until nine at night, although the surrounding hills ensured that when the sun set below them, darkness descended shortly thereafter. The earth had erupted into life. The cows and sheep dozed under the shade of the trees at the height of the noonday sun. Michael and Bobby were with them. Both horses, but Michael in particular, acted like sheepdogs, guarding the lambs against predators like coyotes, which were plentiful and roamed in packs and would like nothing better than to bring down a lamb or two.

Kyle was baling the hay in the field immediately below the farm. He wore shorts, a baseball cap, and protective gear to protect his ears from the assembly-line din of the machinery. The baler, which, like the tedder, was pulled behind the tractor, was old and worked erratically. Kyle had to stop frequently to adjust the controls on the baler. If the tedder sounded like the engine of a boat, the baler had the monotonous thumping of an oil rig as it caught up the loose hay and compressed it into bales held together by twine. *Thump,* a bale was extruded; *thump,* another bale was extruded; *thump,* and another. Bales lay strewn from one end of the field to another, maybe 160 in all. The teeth of the baler picked up the hay and the plunger pounded it into a square bale and the knotter tied the bale together.

The tedder and baler would become a point of contention between Jennifer and Kyle and the person from whom they borrowed the equipment. The lender had wanted what Jennifer and Kyle felt was a lot of money per bale, but they had negotiated their way around that obstacle. Then the lender asked that they keep track of the number of hours of use, according to a meter on the tractor, but the meter was already broken and so it was an imprecise estimate, which displeased the lender. This would not be an issue if they weren't dependent on the lender for a second loan of the same equipment later in the summer, for the Rowan. The lender already thought that they had had it too long: "Kyle got the mower at noon yesterday, and the person called at two to ask when it was coming back," Jennifer said.

There were phone calls, and then long letters staking out position, and the upshot was, when Jennifer asked about borrowing the equipment again, there was no answer, a silence more eloquent than words could be. This was one obvious disadvantage of dependence on the kindness of strangers: If there was miscommunication or no communication at all, resentments were harbored and associations born of necessity could dissolve as rapidly as they were formed.

A woman who lived down Dearing Road, and with whom they bartered goods, was talking about buying a tractor, but that was months off, maybe a year, if it happened at all. Come August, with no other possibility of equipment anywhere on the horizon, Jennifer and Kyle would decide to let the animals graze the remaining grass; there would be no second cut this year. To make up the difference, they bought bales from a neighbor, Bob Post, a dairy farmer from whom they bought their milk. He gave them a very good price, more than fair: $1.75 per bale, 440 bales in all. They had made 1,820 bales of their own, and bought 440, which brought them up to 2,260 bales, enough to see them through the winter.

Post was contemplating getting out of the business. He was in his seventies, and had left the farm only once in his life, to serve during World War II. This spring, he had suffered a debilitating injury when he had had the misfortune to meet up with the wrong end of a cow. The heifer had been dancing around while he was trying to milk her and had fallen over, pinning Post underneath her, fracturing his leg in two places. He had lain prostrate on the floor of the barn for an hour before anyone found him.

Post was like a lot of old-timers—pride and fear and embarrassment precluded him from asking for help, even basic medical help. Days went by; he had been rendered immobile, his leg was nearly twice its normal size, and yet he stubbornly refused treatment. Only the combined pressure of his family and friends—Jennifer had persuaded a doctor she knew to come take a look at his leg, which looked bruised and blackened and swollen to the point of danger to the limb—convinced him to finally seek medical assistance. Post was afraid, he admitted, that if he went into the hospital, at this point in his life and with his injuries, that he would never come back out again, never see his farm again, which would have been a kind of death. (He died in the spring of 2004.)

July began dry, but then came weeks of rain, torrents of it, and the crops began to droop, not from too much sun, but from the deluge. The rivers were near flood stage, turbulent and muddy, roaring past with a speed and at a volume that was reminiscent of spring, when the melting snowpack and ice brought the water levels up so high that they threatened to spill over onto the roads. First farmers had complained that there wasn't going to be enough rain; now they complained that there was too much.

On an early July day, one of the rare sunny ones they would have that month, Kyle was in the milk house, discussing with Ben and Meg the logistics of getting goods to the Lebanon Farmers' Market, but there was a certain impatience in him that suggested his mind was elsewhere—in this case, with his bees. Kyle had many such projects: his woodworking, his mushrooms, his construction on the house and the barn, and the bees. Jennifer had long-term goals but she tended to attack problems in the here and now, and was easily distracted, with one crisis or opportunity ebbing into another. The house is filthy: clean it up. The stalls needs mucking: do it. Brad needs to get from Point A to Point B, so that Jennifer can go from Point C to Point D: arrange it. Kyle had discrete projects to which he attended at different times, many of which met his criterion for why he returned to farming in the first place: "Working outside and working outside and working outside."

The day was pleasantly warm and sunny, with almost no wind and distant, high clouds that seemed in no hurry to go anywhere. The clover was thick in the pasture and the bees would be swarming. Kyle strode down through the pasture to the corner of field where the four hives stood, carrying with him protective headgear, a smoker, and various tools. There were four hives in all, painted white. One colony of bees—the queen, the workers, the drones, and the larvae—lived in each hive. Each hive was comprised of boxes and frames; the boxes, called "supers," looked like stacked drawers. Within each box were frames, in which the bees nested. Somewhere in that network of frames was the queen bee. Kyle had originally purchased the bees as "nukes," the shorthand for a nucleus colony: the queen, and several frames of bees and larvae. His purpose that morning was to infuse the hives with a homemade mixture of sugar and vegetable shortening, which would help to keep down mites dangerous to bees.

With the permission of the Morse family, he had started keeping bees on the farm as early as 1997, before they had even closed on the property; a strategem born of the fact that although they were only living a mile away in the trailer, had he started the bees on that property, they would have migrated the short distance back to it once they were moved to the farm. Bees have to be moved long distances from the original hive or else they will always return to it.

The nukes had been acquired in 2002 from an apiary in Vermont, and following that rough winter, in which a number of colonies had perished, he had bought additional nukes from a New Hampshire apiary. The bee population, which hibernated during the winter months, had suffered heavy losses. "This past winter was bad because of the cold," Kyle said, as he began gingerly pulling out the frames, "and there was no winter thaw, which is when the bees clean out the hive. But I think I lost them before that anyway."

Honeybees throughout the Northeast are under attack from scourges of non-native mites, which infiltrate the hives and kill them. There are tracheal mites that infest the bees themselves; and there are varroa, or external, mites that cling to their bodies and suck the blood from them, weakening and deforming the bees to such a dangerous degree that the entire colony may perish. Both tracheal and varroa mites are the result, to a degree, of overbreeding of honeybees.

"Rudolf Steiner predicted honeybees would be extinct within a hundred years because they were being bred in such a specialized way," said Kyle. "That was almost a hundred years ago."

The hives resembled a miniature O'Hare or JFK or Heathrow, the bees ascending and descending like airplanes, coming and going with regularity. He started up the smoker, which would distract them from his intrusive presence. "As soon as they smell smoke, they start eating honey and it makes it so they don't sting," he said, a behavior that could perhaps be attributed to their fear that the hive was burning down.

After he injected the mite mixture, he carefully opened up the supers and partially pulled out the frames, which teemed with bees and had the familiar look of honeycomb. "They're building the comb out well," he said after a brief inspection. "I always have a hard time finding the queen. I don't want to look too hard because it would disrupt the colony. I try to read other signs." He had, in some instances, resorted to dabbing the queen with Wite-Out, to make it easier to find her. His bees would make on the order of 200 pounds of honey in a season. "Bees make honey to feed themselves over the winter," Kyle said, "and humans have modified it so that bees make extra honey."

This property was conducive to honeybees; they had plenty of pasture and clover, and they also had the basswood tree. "Not very showy flowers," Kyle observed, "but it's a good honey tree." There was also, on the edges of their property, poplar, which is a good pollen tree, and, depending on the year, sumac, goldenrod, and asters, all of which were favored by bees. Kyle collected the raw honey, strained and filtered it, but did not pasteurize it. Raw honey is to pasteurized honey as apple cider is to pasteurized apple juice, Jennifer says; its flavor is deeper and subtler, and thus, highly prized.

Once the smoker was going, the bees virtually ignored Kyle, not even getting terribly close to him, so intent were they on executing their tasks. Kyle watched them raptly. When he first started keeping bees, he would become so involved in watching them work—an interlocking precision and orderliness and industry that bordered on perfection; each bee playing its part to ensure the survival and integrity of the whole—that he would forget what he was supposed to be doing. On such a summer day as this, when the world seemed unusually animated, it was tempting to wax lyrical about the invisible threads that bound bees to trees to flowers to grass, all the unseen forces that conducted the clouds, the sky, the earth at a perfect pitch—this feeling that you were sitting on the minute hand of a universal clock.

But grandiloquence, or hyperbole, was not Kyle's long suit. He was wary of the big pronouncement, perhaps because he had been trained to avoid airy, easy rhetoric, or perhaps because he was so firmly rooted in the demands of the farm, which on some days seemed less a well-oiled machine and more a clanking Rube Goldberg contraption, just a whisker away from chaos.

"Sometimes I think about the larger environmental forces," he said, "but I'm also thinking about why the tractor broke down."

Kyle shows Brad the bees from the hive he'd winterized. The honey from the four hives is sold at the farmers' market in the summer.

When Mark Durkee arrived, on the morning of July 9, in a cool rain and mist, he came in a truck filled with the equipment he would use to slaughter and to dress the two calves he would shoot that day. Ben and Kyle would assist him. He was bearded, dressed in brown coveralls and a blue plaid shirt and a slouchy hat that was pulled comically down around his ears, making him look a little like one of the Seven Dwarves in *Snow White*. There was something ancient in his face that had nothing to do with his age—sixty-seven in a month. His was the kind of seamed, worn, grave face with deep eye sockets that you see on the statues of Old Testament prophets carved into the portals of Gothic cathedrals, or in daguerrotypes from the 1840s or 1850s, in which the sitters had to sit as still as stone for four minutes at a time, with the result that you read in the face, in a way you don't in a modern snapshot, the sum weight of a person's life.

Not a modern face, and after a while it came to you why he looked this way: because he didn't live a modern life in the sense that we think of it. Sure, he drove a truck and he met his wife of forty-two years at a diner, where she was a waitress, and apparently a pretty one at that—"Guys would come in and leave a fifty-cent tip for a ten-cent cup of coffee," he said, holding forth to Ben and Kyle. "I was the only one who didn't leave a tip. I wasn't even a coffee drinker." And the back of the truck had—oddly, given his avowed religiosity—a bumper sticker that read HORN BROKEN, WATCH FOR FINGER. But he was not engaged in what we would think of as modern work. He was doing work that his grandfather and father did, in an area in which his family had lived since the 1700s, and he rarely left the area.

His butchering had made him something of a minor celebrity, the kind of person usually referred to by journalists as the "last of a dying breed." Why, he said, even *The Wall Street Journal* tried to interview him once, but when he asked to read the story before it went to print, because he was suspicious of the media, they refused, and so he refused them. The controversy that arose following the publication of the story about him in the local paper he dismissed with a frown as "silly."

"I never heard a reaction like that before, except in last few years," he said, shaking his head.

Still, journalists approached him, never happier than when they were waxing elegiac over the end of something. The last butcher, the last knacker, the last railroad man, the last farmer—all relics from what is often characterized as an older and simpler time, burnished not only by the human tendency to yearn for the past, but to turn it into something that it never was, simply by eliding inconvenient, uncategorizable fact, and transforming a physically rigorous, demanding, complicated life into the "good old days," halcyon and golden.

Durkee would shoot the animals; unlike Jennifer, who felt slaughtering with a knife was more merciful than with a gun, Durkee didn't cut their throats to kill them, only to bleed them out after they were dead. "We're not here on this earth to torture," he said firmly. When he came down to the field, he told Ben that the cows were too close together. Ben would have to draw away the cows that were not going to be killed with a grain bucket. "I aim just above the eyes and below the poll, the hump on their head," Durkee said.

He waded into their midst. The first calf he would kill was at the end. He looked, he aimed, he shot; the calf flopped onto the ground, four legs pointing skyward. The other cows, which had been milling about, jumped back, not because of the sound, which was no more startling than a cork coming out of a bottle, but because of the sudden movement, the disturbance and the blood that bubbled forth onto the ground after Durkee cut the calf's throat.

The cows jutted their heads forward inquisitively and not a little nervously, their large nostrils flaring as they took in the smell. That done, they moved only a short distance away and returned calmly to eating the grain scattered on the ground. The rain began to sluice the blood away. Flies settled on the calf's face. Durkee tried propping it up, so that he could more easily cut away its hide, but its dead weight pulled it over onto its side: "Oh, you son of a gun," he muttered.

"I try not to butcher on Sunday unless it's an emergency. I believe the Lord wants us to have a day off," he said. "We try to observe the Sabbath." He ruminated further. "Nature tells you there must be something, hmm? God." He looked over to see if this was making a dent, if he'd found a receptive audience. "Animals are made on the same pattern, by one maker. We don't know for sure, but it sure gets you thinking." He nodded wisely.

Butcher Mark Durkee of Chelsea, Vermont, shoots one of two bovines he will skin and quarter. The farm sells beef "on the hoof" at the farm, in addition to packaged organic meat they sell at market.

After he, with Kyle's help, dressed the first calf, he was ready to shoot the second calf, whose mother, Prism, was standing nearby. Prism didn't have Tildy's unruffled temperament or benign mien. She was contrary and the expression in her eye was a bit wild. This might change if, as Jennifer suspected, she were given the kind of undivided attention that a pet would earn, but she was not a pet, she was a dam—a breeder. Prism was half Jersey and half Normande, and mottled looking—chestnut markings with a white face and brown circles around the eyes, so that it looked as if she were wearing glasses.

When Durkee shot her calf, she bellowed, unnerved by the blood and the smell and by the calf's twitching body. "She knows she doesn't have a calf anymore," Durkee said, "but in a day or two she'll forget, unlike humans." As he began to dismember her calf, first removing its skin and then cutting off its hooves and halving it down the middle, and then finally decapitating it, Prism trotted over to smell at it, agitated and confused. Even when the calf had been rendered into meat, she still lingered by its side, and nosed its carcass. Durkee pushed her away. "Git, git," he said. She moved but lingered nearby, letting loose a baffled, braying holler every few minutes.

Jennifer had gone to work at the animal hospital that day, and when she learned that the calf was shot in front of its mother, she was perturbed—not devastated, because the calf would have been killed for meat, regardless—but disturbed. If she had been there, Prism would have been isolated in the barn while the shooting took place, and probably none the wiser. The men—Kyle and Ben and Mark Durkee—simply had not given it any thought, and perhaps such niceties were beside the point, given the inevitable outcome.

When the work was done, Kyle told Durkee that Jennifer had written out a check for him, but that it was up in the house and he had to go get it.

"You going to pay me for the pig?" Durkee asked, his face brightening. He had previously butchered a pig for them, and was owed for it.

"Must be, the check's for a lot more than cutting two steers."

"Oh boy, I hope so!" Durkee said fervently.

Prism continued to bellow for her calf for three days after its slaughter. In August, Kyle and Jennifer would sell her—and two pigs and eight chickens—to a young couple with a baby who had moved up to Randolph from Boston, computer analysts who were now eager to try their hand at organic farming. Jennifer was happy to rid herself of Prism, with whom she had always had a contentious relationship, and she expected that Prism would, at this new place, have the advantage of being the only cow, and thus the beneficiary of more attention.

When Jennifer and Kyle delivered Prism to the couple, in September, the couple gazed around them at the green hills and the fields, mother dandling baby, and talked earnestly about "sustainability" and "organic" and "natural." They had on their faces an expression of infatuation and optimism that Jennifer recognized all too well.

On a Thursday morning at the end of July, Jennifer was frantic with activity, trying to do ten things at once. She was scheduled to leave for West Virginia at five A.M. the next morning with Brad and her sister, Laura, to visit Anna for five days, and she was humming with barely suppressed excitement, although it would be a seventeen-hour drive, straight through. So much rain had made the air heavy and damp. Wisps of cloud hung over the hills behind the farm. One of the geese washed itself in a muddy puddle by the barn, preening and waterproofing herself by reaching back with her beak to an oil gland at the base of her tail, and rubbing the oil into her smooth white feathers. The cicadas, which emerge from the ground every seventeen years, were singing, a vibration that began low, reached a buzzing crescendo, and died away, only to start again a minute later, so that the air seemed to thrum with life.

Ben was there, but not Meg. Her grandfather had died five days earlier and she had gone home for the funeral; Ben would pick her up at the airport this afternoon. Thursdays were the day on which the CSA shareholders came to pick up their goods, and the day Kyle went to the Lebanon farmers' market—a chore that had been relegated to him because no one else wanted to do it.

Using a lathe, Kyle turns a piece of black locust wood to shape a bowl. The black locust was a shallow-rooted tree that fell over in an ice storm. Kyle sells the bowls at an area gallery, the farmers' market, and on the Internet.

Business at the market had been spotty, and because Lebanon fell just two days ahead of the Saturday Norwich market, it seemed like a squandering of their resources, to harvest goods that didn't sell and to make a nearly eighty-mile round trip in a truck that had seen better days.

But Kyle, as it turned out, had an aptitude for what is called "customer relations," probably because he joked around with people, and rarely took offense at any complaints that might come his way, while Jennifer seethed, not so discreetly, when customers, as sometimes happened, complained. He went to market alone and doubled their take, and sold some of his wooden bowls at the same time, which made him happy. He was relaxed when he was by himself at the market, just as he was when he was in his wood shop in the barn, head bent over his lathe, turning the bowls. Solitude had the effect of making him serene.

Jennifer and Kyle, and Ben and Meg, had done the harvesting the day and night before, rather than attempting to do both harvesting and selling on the same day, which would require starting work at dawn. The milk room, where the shares were stored, was fragrant with the scent of herbs—basil, parsley, thyme, tarragon, lavender. The excess rain had affected some of the vegetables; broccoli doesn't like too much moisture and mold had formed on its leaves, and the cabbage was particularly wormy. But it had also brought about astonishing growth in the other vegetables—squash and zucchini and greens that seemed to shoot up a few inches from one day to the next.

Jennifer liked to stroll down to pasture where the crops were, for no other reason than to admire them, to gloat over the prettiness and orderliness and fertility of it all. In spring, summer, and fall, both Jennifer and Kyle would take the occasional break from their chores to walk, alone or together, up and down the fields, or down to the river, to look at the land, and to take pleasure in it. This morning, Jennifer reveled in all that plenty. "Last year, we just had straight rows," Jennifer said. "I loved how Kyle and Ben were so creative about planting." The rows turned corners, were perpendicular to one another, and it was entirely possible, once you had entered into the corn and the tomatoes, to feel that you were in a maze from which it would be difficult to extricate yourself. It was a point of pride that they had done this all themselves, without chemicals, without pesticides, without irrigation, with a maximum of effort but a minimum of hired labor.

Jennifer sat on the outside step to the milk room. She was cleaning picked heads of garlic of their accumulated dust and dirt. Brad was picking his way carefully around the baskets and boxes in the milk room. "I want Daddy, I want him," he whimpered plaintively. "I want chicken noodle soup, Mama."

The phone inside the barn rang, and Jennifer went to answer it.

"Why?" she said, in a loud, shocked voice. There was silence as she listened, and then she hung up. She came back out and sank down to the step, head drooping low, disappointment writ large on her face, near tears. It was a call from the animal hospital, telling her that they had rescinded approval for her taking the Friday off, although she could still have the remaining time for which she'd asked; a splitting of hairs, Jennifer felt, that made it impossible for her—given the travel time to West Virginia—to take the trip at all. "I guess I'm not going to West Virginia," she said in a shaky voice. "A woman quit. And Laura Craft is taking a two-week vacation. They're short-staffed." She rubbed at her eyes. "Anna will be devastated. She just bought a house. She's feeling overwhelmed. And I'm feeling overwhelmed. Kyle and I are not communicating."

A break from each other was what they both needed, she felt. Things had been tense between them. Jennifer had been counting on the trip to see Anna, had been counting the days, and she was determined that, for her own sanity, she would find a way to get there. After some juggling, she arranged to leave with Brad and Laura for West Virginia that Friday evening, after having completed her shift at Country Animal Hospital. This would get them there a day later, and cut short their stay by a day, and decrease the interval between the seventeen-hour drives, but, on balance, Jennifer decided, it was worth it.

Kyle later alluded to the marital tension in passing. "I'm trying to be a good husband," he said gloomily, "which is the hardest thing of all." Sometimes when Kyle was at his most serious, he would laugh a little, and his voice would break at the edges of words, as it did now, as if to undercut

Ben helps Kyle remove Remay covering from kohlrabi. The covering keeps insects away from the plants and elevates temperatures on cool days.

the tension, or to minimize the import of what he was trying to convey. "What!? You're not trying to actually be serious?" was something he often said in response to an earnest question.

What passed between them, stayed between them, but the demands of the job—jobs, actually, when you counted the work off the farm—were such that eighteen-hour days, seven days a week, were not out of the ordinary this time of year, during high summer. In fact, they were the norm.

A visitor would look at their pastures and crops and animals and marvel at the immense fecundity and productivity; what nature was capable of producing and what humans were capable of harnessing. But, at this time of year, Jennifer and Kyle saw dozens of rows to be weeded, thousands of pests to be removed, and animals to be herded from one point to another, or worse, chased down from the scattered corners of the farm or neighbors' property. They saw 1,800 bales of hay to be loaded onto the backs of two pickup trucks, one of them borrowed, that could each hold only fifty bales at a time—which meant, roughly calculated, at least thirty-six bumping, knocking, sweaty trips in the old pickup down the road to the barn and back again.

And then there was the harvesting and weeding. That fecund green? Jennifer called it a "hot, hot green," the sight of which made her tense, because of the work it implied. There were times of year she loved on the farm—the fragile, translucent green of early spring, the golden amber of Indian summer, when their growing season had slowed to a crawl—but this was not one of them. Neither was it Kyle's. In high summer they were surrounded by beauty, and almost could not see it, because of the sheer physical toil.

What seemed like a pleasant way to pass the time at first—outdoors, in the sun, the kind of communion with the land extolled by books—lost its charm when you were stooped over for hours at a time picking, say, radishes by the hundreds. You squatted long enough, and you began to feel it in your thighs and calves. You stood up to do it, and you felt it in your back. You settled for a position that was somewhere between squatting and standing, and your neck began to ache.

The sun was high and you began to sweat, and if you didn't bring water with you, which was the prudent thing to do when you were in the sun for three or four or five hours straight, your tongue—not something to which you normally gave much thought—began to seem like a weird and bulky obstruction that had no business being there, and the inside of your mouth became arid and bitter, and your eyes began to sting and wince from the white glare of the sun.

On days like that, that began at six and ended at nine or ten at night, with harvesting and haying or baling going all day, even the normally resolute and stoic Ben became snappish. Bone-weary was what it felt like. Exhaustion set in, tempers flared, moods were sullen, and marriage came to seem an uneasy mixture of resentment and dependence. Talking things through might have helped, but that kind of time was in short supply now, a luxury in which they could not indulge.

Jennifer said "day" and Kyle said "night," and then they argued about which one was right; rarely did they give ground or concede a point, although if it came to it, Kyle seemed to yield before Jennifer did. Jennifer maintained that Kyle became depressed and moody in the summer, because of the workload, and that, in his fatigue, he sporadically crossed a line from impatience to meanness. Kyle intimated, through his behavior and expression (although he rarely said so directly), that Jennifer demanded too much of him, that he was continually being asked to pass tests, the criteria of which he was only dimly aware, and that the standards by which he was measured were sometimes impossibly high, so that he could not but fail them or disappoint somehow.

It was not unusual for Jennifer to say one thing about a particular situation on the farm, and for Kyle to argue the complete antithesis, without either being aware that they had done so. When the raccoon was stalking the chickens, Jennifer painted a portrait of terrorized birds finding refuge where they could, in parts of the barn where they knew the raccoon would not stray—the haymow or next to the lights. To Kyle, the chickens had always sought refuge in the haymow or by the lights; the raccoon hadn't changed their behavior one iota.

Then there was Michael. If a stranger went down into the pasture where the sheep were, Michael would crowd and bump the stranger, get between the intruder and the sheep, snort loudly through his nostrils. According to Kyle, this was a classic warning-off strategem of intimidation, and one that worked well. Oh no, Jennifer would say. That's just Michael. He wants attention, he always does that.

Borrowing a soil-sampling auger from his dad, Brad chops brassica on a morning of gathering crops for the CSA subscribers and the Lebanon Farmers' Market.

When they talked about the future of the farm in five years, "ifs" began to creep in. If they were still married, if they were still farming, if they were still living here.

"We've always had a hard time working together," Jennifer would concede later in the fall.

What couple doesn't argue? What couple doesn't think, at some point, that they have made a terrible, terrible mistake? What couple doesn't think at some point that divorce might, after all, be the answer? Did they really think that their problems communicating were worse than anyone else's?

Kyle made a face. "Yes," he groaned.

"We just have different styles of doing things," Jennifer maintained. "And we both don't listen to each other's priorities," she admitted.

Would they ever quit farming?

"Never," Jennifer said. "Never."

Kyle was silent.

But there was a child to consider, a son to whom they both clung, who bound them together, and to whom they turned for succor. "If something were to happen to Brad, I would go mad," Jennifer said. Kyle rarely spoke in such dire terms, but he was alert to Brad's every move—his presence, his absence—and as protective as a mother bear with a cub.

Marriage was imperfect, infuriating, chafing—but what was the alternative? Giving up the farm? Separation? Moving elsewhere? There was an unsung heroism in keeping a family, *this* family, together. They had come too far down this road to turn around, and even if they did, the way back would likely seem as long and difficult.

The following Thursday, August 7, Jennifer was back home; Anna and Rick had returned north with her, and were staying on the farm before heading over to Jennifer and Anna's parents, in East Middlebury. In Jennifer's absence, Kyle, Meg, and Ben had looked after the farm. (Solange had finished her apprenticeship and returned to France.)

While Jennifer had been gone, Ben had come down with stomach flu, and had spent a day and a half lying on the couch in the front room, feverish and vomiting. Whatever it was, Meg hadn't contracted it. There appeared to have been miscommunication between Kyle and Ben and Meg. On her return, Kyle had reported to Jennifer that Ben and Meg were grumpy, exhausted, at the breaking point—they had hit the wall that at some stage all apprentices seemed to reach, when the novelty has worn off and all that is left is work and more work. They hadn't wanted to work on Sunday. Normally, that would have been their day off, but with Jennifer gone, Kyle had felt that they should pitch in.

"I'd asked them if they could help bring in lambs on Sunday. I'd asked them a special favor. He, Kyle, thought they were going to *work* on Sunday," Jennifer said, with a touch of exasperation. "Kyle had reported to me that they were being mutinous, refusing to do what he asked. I said, no, that Ben and Meg and I had discussed what was to be done. I'd made out this list, all the things that I didn't know whether Kyle knew to do. Ben and Meg and I talked about what to put on the list. Kyle said they wouldn't do anything that was not on the list. That's because we talked about what to do."

Thursday was one of the busiest days of the week. The shareholders would begin arriving after 11 A.M. to pick up their food; the Lebanon Farmers' Market opened at around lunchtime, and the Norwich Farmers' Market was two days away. All three enterprises entailed picking pounds of produce, moving it from the fields to the milk house, cleaning and bunching it and then putting it into baskets for transport to the farmers' markets, or for pickup by the CSA shareholders on the farm. (The produce was arranged in baskets in the milk house, and the shareholders assembled their own bags of produce, according to whether they had full or half shares.)

Because of this crush of activity, Jennifer was marshaling her forces. Rick would take a load of produce to the South Royalton market, which had called in an order. Kyle would go to the Lebanon Farmers' Market. Ben and Meg would finish cleaning and preparing the produce for CSA pickup. Ben wanted to work Bobby, and that needed to be scheduled in somehow. Jennifer would enlist Meg and Anna to help her move the sheep into the barn. In addition to culling sheep for slaughter, she would check them for two possible afflictions endemic to sheep: foot rot, a bacterial infection that can lead to a terribly painful, debilitating wasting disease, and sun scald, an irritation of the foot (similar in feeling, Jennifer said, to walking barefoot on hot sand or rocks), that comes in the kind of wet conditions they'd had on the farm this month. Kyle walked into the milk house. The truck was backed up to the door, to load produce.

"Do we have a pretty good load of basil for market?" he asked.

"None," Meg said.

"I need some," Kyle said.

"Take a little out of the baskets in the basement," Jennifer said. "Did you look in the baskets there?"

"They're crappy," Meg replied. The basil on the top of the baskets looked rotted, the leaves darkened and spoiled.

"Not underneath," Jennifer said quickly.

Jennifer looked over the assembled baskets of produce: gleaming peppers, golden beets, perfect, glossy cucumbers.

"We don't have filet beans, but Ben's going to pick those, yes?"

"I think so, but I don't think he knows that," Meg said doubtfully. "Ben wants to work Bob today."

"If he picks beans," Jennifer said firmly, "it'll take no more than three hours."

"I hope not," Meg said, and sighed.

"We'll be checking sheep's feet for foot rot," Jennifer said. "Bad, bad stuff. It's serious. But I think it's sun scald. We can bring lambs to market on Monday. Just bring 'em in and leave 'em in so we don't have to go through all this again."

Ben walked into the milk house. Jennifer buttonholed him immediately. "Ben, can you be on filet bean duty this morning and do Bob this afternoon?" Although couched as a request, it was really an order. Ben hesitated. He didn't say anything, but this was clearly not how he wanted things to go; his heart was set on the horse, not the harvesting. His face shifted subtly from agreeable to tense, and he merely nodded to Jennifer's rattled-off requests, said "Yup. Yup. Yup."

"Are you guys writing down what the NOFA woman will want to know?" Jennifer asked them, apropos of the inspection scheduled for three weeks later, on August 22. She seemed galvanized by her time away; a woman who normally exuded more ebullience and stamina than most, she now appeared to be radiating purpose and intensity at a level that approached preternatural. Rick appeared on the step of the milk house, carrying a copy of a book by Isabel Allende, and she asked him to take an order of vegetables to South Royalton. "So, Rick, you know where SoRo market is?" He ran through the directions for her, and she nodded. "Are these ready to be loaded?" he asked, looking at crates of vegetables. "Yes," she replied.

Meg finished weighing produce: "We have nineteen and three-quarters pounds of tomatoes," she said triumphantly. "Whoo-hoo," Jennifer yelped—$96.80 in produce, to be exact, with beans, onions, pattypan squash, beets, tomatoes, squash, cucumbers, and a second delivery of corn and potatoes to follow in two days. Not a princely sum, but of such small sums is a living earned.

That concluded, Jennifer walked down to the pasture just below the barn. She planned to move fence around so that she might, in turn, move the sheep to fresh grass. Seeing her, the sheep sensed what she was about to do, and their baah-ing took on a particularly piercing and urgent quality. "They're warning me they're restless," she observed, looking at them lined up along the fence, dutifully but impatiently. "They got out while I was away." She hustled up to the barn to turn off the electric fence, and hustled back. She could either let them through onto new grass now, or watch helplessly as they burst through on their own, scattering far and wide. In that case, it would take hours to round them up. The fence came down and the sheep galloped toward her. As they ran, she cast a quick, expert eye over the ewes, spotting one that was trailing behind the others. "That thing is dragging its leg. That leg is broken. We'll ship it on Monday," she said briskly. She stopped another one and picked up its hooves and sniffed. "Probably sun scald, burned feet, and with this rain, it's a nice little environment for infection. Three years ago, when it was so rainy, some of the sheep got tapeworm." She cast a further, assessing eye over her domain: "Wouldn't you know it?" she said, looking at the pond, which was slopping over with water, "the year we get the irrigation up, we don't need it." (They had dug the well the previous fall in order to mitigate against drought; because of this spring and summer's rainfall, they would not have to use the well after all.)

The sheep were happily ensconced in new grass, and Jennifer continued to move fence, folding it up as if it were tennis netting, so that she could reposition it across the dirt path that led from Morse Road down to the pasture. "Ben and Meg have been great, but the chances of them coming back are slim to none. They're ready to do their own thing. They're way beyond being apprentices. I'd love it if they came back, but I doubt they will. Which means we're really going to have to think hard about next year's production."

Meg and Anna came down to help her round up the sheep. Jennifer held the all-important grain bucket. Meg chased a stray lamb that was not walking properly, got hold of it, and Jennifer examined it. Its hooves were badly overgrown, as human toenails might be if they were never cut, which had led to obvious difficulty in moving. "It's not foot rot," Jennifer said. "It's sun scald. Foot rot smells." Their sheep had never had foot rot. But because they had three new ewes, acquired from farms in New England, she had to be vigilant; once foot rot gets into a flock, it takes rigorous treatment to get rid of it.

With clippers, she cut the hoof back and cleaned it and sent the lamb on its way. Another lamb was more seriously injured. They could see that it was dragging one leg behind it. This lamb—#78 yellow—was the offspring of one of Jennifer's favorite ewes, a sheep named Charlemagne. Somehow a wire—what looked like netting that had come loose—had become wound around the bottom quarter of its back left leg. The wire was so deeply embedded that it had cut down to the bone, and the flesh and skin had now swollen around it. It was questionable whether, first, they could remove a wire so deep in the leg and, second, if they could remove it, whether the lamb would recover use of its leg. If not, it would be shipped to the slaughterhouse.

They managed to get the lamb up to the doorway of the barn. Meg held it so that it was sitting up, its back to Meg, legs sticking out. Jennifer got out the wire cutters and Anna looked over Jennifer's shoulder. Very gingerly, Jennifer began to cut away the wire. Both she and Anna felt the leg; it was warm, which meant that the blood was still circulating and the leg hadn't lost all feeling. The lamb stood a chance then. Jennifer carefully, carefully unwound the wire. The wound was deep—you could see beneath the pink flesh to the white bone—but clean. No infection around it, only the swelling that had come from the insult to the area. Jennifer sprayed the leg with gentian violet, a bright purple disinfectant that is permitted by NOFA. She let the lamb go and it hopped down the slope, damaged leg in the air. Everyone watched it to see whether it would put its leg down.

"Don't you want to put her inside?" Anna asked, concerned.

"No," Jennifer said.

The lamb was moving along quite speedily, but still with the leg in the air.

"Sis, don't you want to put her inside?" Anna asked again.

"No," Jennifer said in a tone that permitted no dissent. "She should be with her mother."

A minute later, the lamb touched the damaged leg tentatively to ground, and picked it up again. It tried again a few minutes after that. A few days later, the leg was healing, and the lamb was trotting along as if nothing had ever happened.

The NOFA inspection was an annual rite of passage, typically occurring in the summer so that an inspector could see a farm at the height of production. The inspection was, in a sense, a summation and validation of everything Jennifer and Kyle had learned and put into practice as farmers: the organic methods brought to fruition, a yardstick against which they could measure production and methodology. There had been improvement upon improvement, each one built on the back of the one before.

The soil was, five years on, noticeably more friable. It crumbled easily between the fingers, was richer-looking and -smelling. It was still predominantly clay, and clay is heavy, but it didn't resist them the way it had when they first plowed it up. Where there had been burdock and goldenrod and aster in the pastures, were now clover and timothy and dandelion and other grasses more suited

A chick, named Zebra for the stripes on its back, looks for its mother in the grass by the barn. Her mother had laid her eggs in the hay in the Horse Barn, and hatched six chicks.

for grazing by both the cows and the sheep. The riverbank was fenced and trees planted along its length, to reduce erosion. Livestock were no longer permitted to graze the steep wooded embankment going down to the river, which reduced erosion and the grazing of seedling trees. Not all of these changes were germane to the NOFA inspection, but they were significant, and provided a framework for the overall restoration of the property.

In previous years they had had an inspector who subjected them to a rigorous accounting, followed every paper trail, noted every purchase, queried every decision. This year, a new inspector was promised, and there was some tension about what kind of inspector she would be. She was due to arrive on the morning of August 22. Kyle was there to talk to her; Jennifer, who was at work, would join them later.

The inspector, Jean Richardson, was in her fifties, with short white hair, glasses, pink cheeks, and the faint vestiges of an English accent—not the clipped, plummy tones of the South but the softer, rounder burr of the North—Northumberland, to be exact. Her manner was soft-spoken and tranquil, disposed, it seemed, to take their side, not to subject them to inquisition, although the questioning was certainly detailed.

"The map of the farm is correct?" she asked, sitting at the table near the kitchen. Kyle sat next to her, chin propped on hand, elbow on table, posture slightly slumped. From his angle, he could glance at the NOFA file she had in hand. He was attentive and calm but despite the outwardly casual posture, his leg under the table jittered away.

"It's unchanged? All the fields have been free from prohibited substances for at least three years?" She peered over her glasses at the file.

"From last year, the only thing in your letter," she said, looking back and forth between papers, "was to track the organic practices of your neighbor, regarding use of Greenhouse Number Two."

"We've had another change," Kyle explained. "We're not using that greenhouse anymore. Greenhouse Number One never turned into a greenhouse. It's a chicken house. Greenhouse Number Three hasn't been used for a couple of years."

"What's the situation with Greenhouse Number Two?"

"We're not using it," Kyle said.

"Your plans for Greenhouse Number Two?"

"We're not using it," Kyle reiterated. "The woman who owns it took it back." This segued into discussion about which greenhouses they were using this year, if any, and where they were, and what they were like. Kyle told Jean Richardson that, this winter, they had used for the first time a greenhouse in South Randolph. She delved further, into its history and layout, whether it had been used for organic materials, and whether the wood with which it had been built was pressure-treated. It was organic, yes, Kyle told her, but the question of the wood remained open.

How much land did they farm, Richardson asked? Land in use, Kyle said, was 13 acres on the farm, and 31 total leased acreage; 44 acres all together. "All farmed organically?" Richardson asked. Yes, Kyle answered.

"We're still seeking certification of hay, vegetables, garlic, and livestock," Richardson added. She went over the farm's record: what did they use to treat the soil, if anything? Did they do soil tests? Where did they get their seeds? Were any of them treated? How did they control pests? If they used insecticides, what kind? There was some discussion over the use of the B.t. *(Bacillus thurengensis)* on the Colorado potato beetle.

Richardson frowned slightly. "We have to check on the B.t." She began leafing through the NOFA manual to see what the standard was on its use; she believed it was prohibited. (In fact, Jennifer had bought the B.t. through NOFA; its use was allowed, as she confirmed later to Richardson.)

"I knew there was a problem but I didn't save the container. We used it all up," Kyle said.

"B.t. is a challenge for the National Organic Standards Board," said Richardson. "The challenge is to determine what substances are in it because different companies produce different brands.

"How do you control weeds?" she continued.

"Hand pulling and weed hoes," Kyle said. "In the pasture, we're digging up bull thistles to get the weeds out. In the hayfields, it's not a problem. Also we covercrop and do a lot of mulching."

"Do you do crop rotation?"

"Yes."

"Describe it."

"We rotate each family of vegetables so nothing is in the same place for a three-year rotation."

"Are the fields well isolated from potential contamination?"

"If, by isolated, you mean the river?"

"Fifty feet?"

"Yes, the river."

"You've planted buckwheat?" (Buckwheat is an effective cover crop, a way of keeping down weeds.)

"We've finally been hammered by galinsoga," Kyle said. (Galinsoga is a particularly noxious and invasive weed.)

"We've got fusarium and nematodes listed here," Richardson said.

"Mostly on the beans."

"You're just letting it happen?"

"Yup, we like to find reasons not to pick beans. Right, Meg?" Kyle said. Meg had just come into the kitchen to get herself a glass of water. Her face was flushed from the heat. "Yup," Meg answered brightly and vaguely. Richardson asked whether the carrots had suffered an infestation of nematodes.

"Not really," Kyle said. "We've gotten ugly carrots, though. Could be nematodes."

The questioning continued. Was the hay stored on the farm? Yes? Was there any off-farm storage? No. Where were the vegetables stored? Milk house. Was the hay wrapped in plastic? No; square bales were kept up in the haymow. Was the water irrigation from a drilled well? No water tests yet? Right, Kyle said, with a bit of an edge in his voice. Planning on it? Yes.

Richardson informed him that she would need to see receipts and production logs for the produce, at which point a look of unwelcome surprise crossed Kyle's deliberately neutral face. His eyes widened. "We don't have production logs for vegetables."

"It's not usual," Richardson said serenely. "Don't worry."

"Good, because you scared me," he blurted out. "For things that go to stores and farmers' markets, we have records."

"Now the beasties," Richardson declared. They went over how the animals were raised and sold, how many of them, how they were tagged, what was federally inspected and what was not; what Richardson called tracking the livestock "cradle to grave." Kyle answered quickly and automatically, but his patience was wearing thin, not because the questions were unexpected or unwarranted or confrontational but because it was a drawn-out, detailed interrogation that, by its very nature, put one on the defensive.

In his speech and manner, Kyle was more formal than he usually was, terse, as if he were afraid of giving the wrong answer or as if, by elaborating too much, he would unwittingly betray some oversight or failure on their part. When Jennifer breezed in, all smiles and affability, he was visibly relieved to see her—"Heeeere's Jennifer!" he said with a smile—and turned the proceedings over to her with alacrity.

Richardson looked at Jennifer closely. "Are you one of my former students?" They examined each other. Yes, it turned out, Jennifer had, in all likelihood, taken an environmental studies class from Richardson, who had taught at UVM for twenty-one years.

Jennifer and Jean Richardson went over all the animal records. Jennifer treated her as a colleague, one with whom she could collaborate. They did a walk-through of the barn. Richardson, who had raised sheep herself, went over the finer points of husbandry with Jennifer. Three weeks later, they received their certification, with the proviso that they attend more closely to the treated wood at the greenhouse. The report, Jennifer says, was "glowing."

"She said what a good farmer I am."

"*Farmers*," Kyle corrected her.

The Tunbridge Fair advertises itself as a World's Fair, and has been doing so since 1875. The world really doesn't have anything to do with it, although visitors come from all over the state during the four days—from a Thursday through Sunday—that it is held annually in September. The fair has been going since 1867 on the same 32 acres of land in a river valley off Vermont Route 110, where it abuts the narrow, meandering First Branch of the White River and is set off by a succession of steep, heavily wooded hills to the east and the west.

There are still farms the length and breadth of the valley, which runs some 30 miles north from South Royalton to the granite quarries at Barre. Only two events have ever preempted the fair—the 1918 Spanish influenza pandemic, and World War II, because of the numbers of men called to service. There was discussion on September 11, 2001, whether it would go ahead as planned on Thursday, September 13, 2001. It did.

The fair is more of a family affair now, less the occasion for rambunctiousness and public drunkenness than it used to be, although there is a demolition derby and a potbellied pig race and a beer tent that now strictly cards would-be drinkers. No more girlie shows with the strippers going wanly through their paces, no more Pig Baby, Cardiff Giant, World's Largest Rat, or Fire-Breathing Dragon from Sumatra, no more inebriated fistfights, no more gambling. Sober types are no longer booted off the grounds at three P.M., so that carousing and revelry can have their day without interference.

There is no better place, in a sense, to see the old Vermont and the new Vermont rub elbows and jostle up against each other—the hill people, enormous, bearded, rough-looking men who look like something out of an Albert Bierstadt or George Caleb Bingham painting of nineteenth-century mountain men or river traders, wary of too much civilization, and the women who accompany them, often as big and rough looking, with hair hanging to their waists. The wizened older farmers whose skin is mottled by age spots and sun, with the nasal, elongated vowels of northern New England. The fresh-faced boys and girls with their prize oxen and heifers, taking their duties to heart. The straight-as-sticks cadets from Norwich University, the military academy in Northfield, Vermont, who, wearing fatigues, help to park cars and oversee admission. The bored carnies, now predominantly Latino, who admit you to the various rides. The adolescents, roaming in packs. The concession stand overseers hawking absurdly large stuffed pink bears if you'll just pay them a dollar to take your best shot at the ducks in the shooting gallery. The yuppies squiring children, carting with them the requisite Baby Bjorns, Mountain Face backpacks, Polar Fleece pullovers, Merrell moccasins, and Graco strollers. The white-haired church ladies who arrange with tremendous care the homemade quilts and flowers in the exhibition hall.

The agricultural exhibits—the cows, the oxen, the sheep, the pigs, the rabbits, the geese, the chickens. The pungent smell of still-warm manure. The midway, teeming with people, day or night. The fried dough, the dumplings, the ice cream, the french fries, the hot dogs, the hamburgers, the burritos, the fruit smoothies. The shooting galleries, the merry-go-round, the Ferris wheel, the bumper cars. The oxen pull, the sulky races, the butter churning, and the living history exhibits. Hundreds of cars and trucks parked in the fields, glittering in the sun. A blur of red, blue, orange, and green neon and jangling, hurdy-gurdy noise. For four days each year, a virtual city arises out of the ground here, with all the drama and life and people and allure of a metropolis.

Jennifer and Kyle had brought Brad for the first day of the fair, a Thursday, that traditionally marks the start of festivities and is set aside particularly for families and children. Jennifer's parents, Louis and Beverly Megyesi, had come along as well.

Kyle carried Brad on his shoulders, thick arms resting easily on Brad's skinny legs. Brad insisted on trying nearly every kiddie ride, rode the horse on the carousel without help, waving enthusiastically to his mother and father. As he passed them by, he could see, if he looked closely, faces suffused with pride and tenderness and perhaps not a little fear to see him wavering alone atop the prancing horse. Look what they have wrought! A son!

Ben and Meg would turn up later. This was their last hurrah; they were leaving the farm for good on Monday, to return to Ben's family's property in Massachusetts, where they would find jobs over the fall and winter and try to save money for a place of their own.

This year, opening day happened to fall on the second anniversary of the September 11, 2001, terrorist attacks on New York and Washington. There would be, an announcer said over the booming

Brad and his dad Kyle look at the midway at the World's Fair in Tunbridge, Vermont. The fair, a big family outing, is close to Fat Rooster Farm. Its arrival marks the time of year when activity is beginning to slow down at the farm.

Jennifer looks at the kale entries judged at the Tunbridge World's Fair. Local organic farms do well in the competition, possibly due to the variety of crops they grow.

loudspeaker system, a moment of silence at seven P.M.—a time that had nothing to do with the actual time the attacks took place, before nine A.M., but a time at which the fair was crowded—and at exactly seven P.M., not one minute before and not one minute later, the voice came back and asked visitors to stand where they were and observe two minutes of silence. And for two minutes, astonishingly, the fair was completely and utterly still, with hundreds of men and women and children frozen in their tracks, and the rides whirred to a halt, so that the air seemed unnaturally quiet. Kyle held Brad on his shoulders and when Brad tried to talk, Kyle shushed him gently. Jennifer stood next to them.

Jennifer and Kyle had entered almost seventy-five vegetables and canned goods into the judging contest, held every year in the Dodge-Gilman building, a spartan-looking hall not far from the agricultural buildings. Pumpkins, tomatoes, eggplant, peppers, corn, eggs, onions, garlic, leeks, radish, floral arrangements, relish, rhubarb, chard . . . the list went on and on. These contests were taken very seriously. *Very seriously.*

There was, for example, a great stir going on in the front of the hall, where two elderly men were weighing pumpkins on a large scale: 300 pounds, 400 pounds, 500 pounds; pumpkins so bloated and huge and sprawling that it took as many as six men to move one 500-pounder from the back of a pickup truck onto the scale, where its enormous weight drew impressed murmurs and loud joshing. Kyle's 200-plus-pounds pumpkin was a pipsqueak by comparison, but one of his Warty Tan Cheese pumpkins—an ugly name for an endearingly ugly, lumpish pumpkin that looks as though it is afflicted with elephantiasis—was slapped with a blue ribbon in the Weird and Strange category, and one of their zucchinis came in second in the Largest Zucchini category—hotly contested races all.

"Jennifer is disgustingly competitive," Kyle joked, watching her as she raced from entry to entry, making the occasional joking, dismissive remark about other prizewinners and, to be fair, issuing compliments on produce that was particularly lustrous, particularly perfect. Both, though, seemed to take an almost childlike pleasure in finding blue, red, or white ribbons attached to their entries, as if they had been awarded gold stars for good behavior in the second-grade homeroom, and took care to point them out to Brad.

A first for leeks! A first for radish! A first for an arrangement of calendulas in a vase! A first to Kyle for a wooden bowl in the Crafts category! A Grand Prize in the Best of Swine category! And, Jennifer squealed with excitement, a first for best overall collection of canned goods, six jars in all; a prestigious win in a coveted category.

These were the fruits of the harvest, jar after gleaming jar—summer caught and preserved at its zenith. All that industry and domesticity put to the service of getting through the winters; a largely rural skill thought to have gone the way of the milk wagon, but that will still thrive as long as people have kitchen gardens. There was something hopeful and poignant about seeing the jars of produce lined up on the shelves, one after the other, with the assorted red, white, and blue ribbons tacked to them. Hopeful because they implied that there would be a next summer, with more produce to be harvested, and a summer after that. Poignant because these agricultural fairs came in early fall, when there may have already been a first killing frost, or it is not far away, and the fruits and vegetables and flowers had turned the corner from middle age to senescence, slowly and inevitably dying off.

This year the frost would not arrive until the end of September. This was not as late as 2002, when frost didn't come until October, which seemed alarmingly late, as if the season wasn't behaving as people expected it to. The days had the warm golden haze of Indian summer and to the west, toward the Green Mountains, and beyond them, the Adirondack Mountains, the hills shimmered in the clear light, but the night air had a bite to it, and the darkness settled in a little earlier each evening.

In the morning there was likely to be a hoar frost—an overlay of heavy white dew on the grass that evaporated as soon as the sun took hold, but which returned the next evening. The trees that were not so healthy had already begun to turn color at the end of August—maples that were green on one side, and red on the other, or mottled gold and green. In a few weeks, the hillsides would be awash in great bands of color—swaths of red and gold and amber and burnt orange. A few weeks after that, the leaves would begin to flutter down in earnest. One good wind at the end of October or early November would strip the trees of their leaves almost overnight, leaving the trees startlingly bare and bereft.

People started to make dark noises about winter coming, the end of light and warmth, and the beginning of cold and privation, when the land shrank back into itself and the humans retreated to their houses and the animals began to seek shelter. It was a perverse thing—and quite Yankee, a vestige of the Puritans' Calvinist outlook—the way people, basking in September warmth weeks before the snow came, began to lower expectations and anticipate trouble and make unhappy predictions, as if pleasure was not an end in itself but a foolish indulgence, an open invitation to hard times and harder weather.

A week after the fair, Jennifer was out in the fields picking through the vegetables. "The tomatoes are done," she said. There was still fruit on the plants, but the leaves and stalks were now withered and blackened. They still had squash, beets, carrots, chard, brussels sprouts, cilantro, dill, parsley, kale, and onions in abundance. Jennifer stepped into the cucumber patch, where the cucumbers, now past their peak, had become so large and heavy that they had separated of their own accord from their stalks and had begun to rot.

"These infernal cucumbers," Jennifer said, lobbing one after the other of the monsters over the fence to the cows.

A few more weeks, and Jennifer and Kyle would let down the fence that had kept the livestock from getting at the crops, and within two hours, the remnants of the crops would be gone, eaten, as if they had never been there. The onions and leeks and shallots and potatoes were done, all sold. The honey had been collected. The garlic had been gathered in July and hung up to dry in the barn; it would go to market soon and they would have to plant new cloves in October.

The CSA shares were winding down, for which both Jennifer and Kyle were grateful, but particularly Jennifer, who did not take kindly to some complaints, reacting as if her children were being criticized, not her vegetables: "We wish the produce could have been cleaner," as if their vegetables should have looked like store-bought produce, instead of having come out of the earth; or, "Why didn't we get lamb in August?"

Minor complaints, really, but all of which seemed to belie the fervor with which some of the shareholders donned the organic mantle. Perhaps these complaints were an indication that while some of the public liked to think they were ardent supporters of the organic ethos, in reality what they really wanted was what they were already used to in the supermarkets, and were suspicious of anything different.

The lambs were gone, the ducks were gone, the meat chickens too. Out of the fifty turkeys they had bought, there are only fifteen left for slaughter at Thanksgiving. Jennifer and Kyle had hoped that they would bring in substantial income. After the losses they had sustained from accident and predation, however, the turkeys were judged to have been a "dismal failure." Because of the losses, Jennifer and Kyle would be some $2,000 short—$2,000 that could help sustain them through lean times in January and February.

The labors of summer were akin to the spent passions of lovemaking: so much sound and fury and conviction, now sated and exhausted. Jennifer looked drained. "We're still getting three hundred dollars' worth of orders from the co-op, but there's no one to bring it in. Last week was horrible. It's tougher without Ben and Meg."

Between Jennifer and Kyle, just the two of them now, they'd had to count all the sheep and cull some for slaughter. The cows had gotten out. She and Kyle had been arguing. The pigs that were butchered by Mark Durkee at the end of August had gone to a new custom meat cutter, but the order had been screwed up somehow. They were still obligated to go to the farmers' markets in Norwich and in Lebanon—once they'd signed on, Jennifer and Kyle felt, they couldn't drop out; their customers counted on them—but there was no one to help them, and they still had their jobs off the farm to go to.

They'd had a bull for a time which they'd rented from a couple they knew, Patty and George Gast. For a bull, Buddy was very gentle. He had done what he was supposed to do—impregnated

Jennifer picks dill from among the celery and buckwheat for one of the last Saturday farmers' markets.

some of the cows—and could be easily led by halter, a kind of Ferdinand the Bull who liked to smell the flowers, not a snorting, ground-pawing behemoth to which you had to give a wide, respectful berth. He, too, was gone. "They took him away and killed him," Jennifer said, a touch mournfully. "I wished I'd had the money to keep him. Yup, Buddy the Bull is hamburger."

Jennifer had told Kyle that the state of the marriage was such that the only reason she stayed was for Brad. And if it didn't improve, or they didn't work together to make it improve, she would leave. It wasn't an idle threat; she had never said this before to him, and he took it seriously. But summers were always hard; each summer seemed to bring its own crisis that had to be weathered—poverty, drought, sickness, unhappiness. Sometimes it seemed as if the only thing missing was a plague of locusts. The amount of work that had to be done drove them nearly into the ground. Attending to a marriage was not that much different than attending to a piece of farm equipment. It broke down, you fixed it, it broke down again, you fixed it again, and on and on. Somehow the repairs were made and the marriage kept working.

Out of the corner of her eye, Jennifer saw that a calf had broken through the fence and was trotting away, shifty-eyed, casting backward glances at Jennifer as if to say, "What are you going to do about it?" Jennifer's face darkened.

"Jerk cow," she yelled, hands on hip. "You shit!" Then she sighed. "Oh, it's not her fault. It was the sheep that got out. When I went to shut the fence, she got out." She would ask Kyle to help her drive the cow back into fenced pasture—occasional difficulties notwithstanding, they relied on each other during these mini-crises—but for now it behaved as a naughty child, trotting not too far away when it saw Jennifer looking at it, halting when it didn't have her attention.

"Things are slowing down," she said. "You know we hired some guy to paint our house? He's charging ten dollars an hour." Although no one had said anything specific to her, Jennifer feared that people would somehow associate the shabbier appearance of the house with their produce. She wanted the property to look well kept, the kind of place that drew you in, not kept you away. The painter was putting primer on the sides of the house, and new paint on the standing seam roof, and the house—the kind of place that realtors call a fixer-upper or advertise as having "character"—was slowly being reclaimed.

In early October, Kyle's parents had driven in three days from Circleville to the farm. It was an impromptu four-day visit, spurred by their desire to see Brad, even though Kyle, Jennifer, and Brad were scheduled to drive out to Circleville the following week for a weeklong stay.

It was the time of year when Jennifer and Kyle could contemplate taking some time away from the farm. Just two days before the Joneses's visit, they had spent two nights at the tony Basin Harbor Club, in Vergennes, Vermont, on Lake Champlain; Jennifer had won the stay in a fund-raising auction on Vermont Public Television. They were taking the week of October 13 to drive to Ohio.

The Joneses had stopped on the way east in Schoharie, New York, the early home of the Defenbaugh clan, to do some genealogical research, which was one of Lois Jones's passions. She was short and a bit heavyset, with gray, short hair and glasses, smiling and loquacious. Sandy Jones was, if such a thing was possible, even more laconic than his son. Between father and son, there was a strong resemblance, in size and in the strong profile (father bearded, son clean-shaven, both with high foreheads, thick hair, and receding hairlines), and in the weighing of words. Both mother and father expressed, in their own polite, tentative ways, their concerns about this thing that Kyle had gotten himself into—farming—a doubtfulness surely colored by their own recollections of difficulties with the Ohio farm.

"We were amazed when he wanted to do this," his mother said. "When he left, he didn't want to have anything to do with it. He didn't want to farm, he liked the outdoors. I never encouraged the children to go into farming, but I didn't tell them what to do either." Of Kyle's decision to shift from a full-time to a part-time job at Marsh-Billings-Rockefeller National Historical Park, she said,

Brad samples brussels sprouts straight from the stalk. The heads gain sweetness after the first frost.

"We hated to see him leave it because it was steady income and health insurance, but he doesn't like the bureaucrats. My husband was not a born farmer, more a forester. My husband knows every tree; Kyle knows every bird."

As to the farm itself, her face screwed up into a moue of uncertainty. "If they love it . . ." she said doubtfully. "They talk like they're successful, and I don't know. I don't check their books. I'm awful glad he has that National Park Service job."

They had given up farming in 1988 or '89, Sandy Jones thought. He had come in from a solitary walk and sat in a chair kitty-cornered from his wife. "No money in it," he said, letting go each word as if dropping stones into water. "My knees were bad."

"You were bowlegged," his wife teased him.

"Not me."

Surely there was something they missed? "Not too much," he said flatly.

In its scope and in what they produced and their organic methods, Fat Rooster Farm was the antithesis of the farm in Circleville. "A whole lot different," Sandy Jones observed. "So much hand work." By contrast, their Ohio crops of soybeans and wheat and corn had been "pretty well mechanized, although primitive to what they're doing today." Fat Rooster was "so labor intensive," which was fine as "long as they can both do it." He scratched his head, and was stirred to something approaching a philosophical disquisition. "People just burn out. It takes a special mentality to be home all the time. I've met dairy farmers whose sons absolutely would not farm."

"We wouldn't want to undertake this kind of farm," he said.

Kyle and Jennifer had been putting off having a meeting about what direction the farm should take in 2004. It was scheduled for one week and that went by, and then another week, and that went by, until it was November, almost Thanksgiving. The day was gray and moody, characteristic November weather when it is not yet cold but autumn has passed, and a creeping melancholy overtakes the landscape. The skies were streaked gray and purple and indigo. The clouds hung low and moved fast, whisking by on the stiffer November winds that shook tree branches and rattled windows and, on the hilltops, sounded like freight trains coming through from a great distance. For the first time in months, there was no foliage to obscure the hills, or the serpentine lines of the rivers and streams. The tamaracks, the last to turn color, flares of gold on the dark hillsides, had shaded to brown. The hardwood trees, now denuded, had the look of toothpicks stuck into the hillsides.

The cornfields had been cut back to a sharp, pointy stubble. Flocks of wild turkeys picked their way through them, looking for grain. The Canada geese had begun migrating south in October. They had massed on the White River and the Connecticut, and on the open, plowed fields, twenty-five to fifty birds at a time, before rising in black clouds, taking to the skies in long, wavering Vs, their determined, emphatic honking carrying for miles until they could be heard no more. There had been a few dustings of snow, a light, grainy cover that vanished quickly; the air before snow smelled vinegary, sharp and cutting, and then the flakes came, imperceptibly at first, dirty gray motes that seemed like dust kicked up from the road before they turned to whirling flurries.

Jennifer and Kyle had picked all the pumpkins and put them on a wagon in front of the house, with a sign saying, BUY LOCAL. Their fields now lay empty, although the sheep and cows were still outside.

For whatever reason, it was clear from the start that neither of them really wanted the meeting that morning. When Kyle came in from his wood shop, he looked grim; the motor on his lathe had broken and he would have to replace it, which entailed, of course, expense, but more to the point, it prevented him from doing the one thing that he felt was truly his own—woodworking. Jennifer's expression, too, was set, as if anticipating trouble, her skin grayish.

"We got a phone call from Killdeer Farms," Jennifer told him. "They're wanting more delicata squash." A safe, neutral topic. They discussed hiring apprentices.

"Although Ben and Meg were really great, we figured out we can't afford to pay that kind of money," Jennifer said. "We paid them $8,300"—a rough estimate that turned out to be incorrect; they were paid more in the neighborhood of $5,500 for the summer.

Covered with a dusting of snow, Cody the border collie keeps an eye on Brad.

"They earned it," Kyle said, and she did not disagree.

"But our biggest seller is lamb, and in terms of that, they didn't do much," she said. Kyle asked her if that was gross profit, or net.

"I think we agreed to do more meat chickens next year," Jennifer continued. "We get calls every day. But no more ducks until I can figure out how to kill them more efficiently. We sold all but four, and we raised fifty."

"I don't want more turkeys," Kyle said.

"Are we having a disagreement about turkeys? We had a horrendous year."

"We always have a disastrous year," Kyle said steadily.

"The average is a ten percent loss," Jennifer said.

"I'm thinking of the year before last," Kyle said. "When they did badly."

"My thing is, it's a lot of money in November when we don't have a lot of other income," Jennifer said.

Silence. She looked at him, and he ignored her and then looked away.

"Are you waiting on me?" Kyle asked. His voice was flat and expressionless.

"I just want more participation," Jennifer said. Her brow was now furrowed, and mouth down-turned. There was more heavy silence.

"I guess we're doing turkeys," he said in a monotone.

"Meat birds and turkeys require different management," Jennifer said.

"I'm not really interested in managing anything *alone.*"

"We did a hundred this year," Jennifer said. "Chickens."

"We can sell whatever we produce, but it's the producing that's the problem."

"It's nineteen hundred dollars a month to produce."

"I don't like doing chickens," Kyle declared.

"How many do you want to do?" Jennifer pressed him. Her voice rose.

He sighed wearily. "I don't know. Pick a number."

"One hundred and fifty."

"Fine."

"I'll check with Luna Bleu, to see if that's okay. It's the same number we had this year, a hundred chickens and fifty ducks." She persevered. "As far as pigs, what do you want to do?"

"I don't understand why you're putting it all on me," Kyle said, with some heat. "We have four sows. That pretty much determines what we do."

"The layers are coming on line in March," Jennifer said. "By fall, another hundred? I don't think we're having a problem selling eggs."

"We have to do a better job of keeping the birds alive," Kyle pointed out. "We need to build a coop."

"We have to figure out where to put them. Where they are now is on a steep slope," Jennifer said.

"I don't know. Where do you want them?" Kyle said, speaking very slowly.

"Could we make a coop out of that blue wagon? How would you move it? Then we'd have to have an electric thing around them," Jennifer said, almost talking to herself.

"Umm-hmm."

They moved on to the leghorn chickens. "We could put them over near Bruce's field," Jennifer said. "But how would we get water to them?"

"The waterline goes up pretty high."

"Over in a corner of the field?"

"I'd like to not detail it all out right now."

"Any other livestock stuff? Sheep?"

"Are we putting sheep up on Charlie's pastures?" Kyle asked.

"That'd be nice, but how would you fence it? How many sheep would we have? Do you want to stay at sixty, or go up to seventy-five?"

"We're, *you're*, never interested in going up to more than sixty," Kyle said. Jennifer pulled out a calculator, and did a tally. "Seventy times one point six lambs per ewe. That's one hundred and twenty lambs! This year we're going to have ninety-six lambs. Where are we going to put them all?"

Organic eggs are ready for washing. The eggs are either sold at the farmers' markets, local co-ops, or to Community Supported Agriculture (CSA) subscribers.

"Then we don't need to grow any more lambs," Kyle said.

Brad was playing with a piece of tape that he had stuck to his nose, oblivious to the tension between his parents. He sidled over to his father. Kyle laughed, and took the tape off.

"I'm interested in Tunis sheep but they're out-of-season breeders," Jennifer continued. Kyle looked surprised. Tunis sheep? Where did that come from? Was this another one of Jennifer's mad schemes?

"I don't like that idea," Kyle said. "Where are you going to get them?"

"A rich couple in New York on the Hudson are giving away free Tunis," Jennifer insisted.

"It sounds sketchy to me. The main thing is, as long as we're not breeding six months a year."

"Should we sell lambs through Vermont Quality Meats or the farmers' markets? We sold thirty-nine lambs at market," Jennifer came back.

Vermont Quality Meats, the cooperative that sold both organically and conventionally raised custom meats to the better restaurants in New York and Boston—places like Babbo, Les Halles, Frank's, Daniel—was having financial difficulties. Kyle and Jennifer had placed lambs with the cooperative, but there had been problems for the cooperative when the restaurants specified exact hanging weights and then the lambs were off by a few pounds.

If the animals didn't come in at the exact weight, the restaurants would not buy them. Which meant money spent on transport and slaughter and dressing, but no sale by way of recompense— a waste. Jennifer and Kyle decided that they were probably better off selling directly themselves, either at market or through custom slaughter. They moved on to the pigs which, at this moment, seemed more trouble than they were worth.

"They sit around and eat, sleep, drink, eat, sleep, drink," Jennifer groused. Their piglets, in particular, had been babied, fed grain and apples and vegetables and milk, with the result that they had 30-pound pigs at eight weeks, when the average is 30 pounds at ten weeks. The lambs, by contrast, were relatively autonomous, not dependent on Kyle and Jennifer for food, fed on mother's milk and grass only. "They're on their moms, outside," Jennifer said.

And, further, they ran into the same problem with pigs that they did with lambs, when they sold to restaurants; specific hanging weights. "It's really hard to deal with specific hanging weights versus a range of weights," Jennifer added. "But VQM is a good backup. The prosciutto pigs are such a gamble. If they don't buy them, who's going to buy a three hundred-pound pig?"

Talk turned to the infamous White Pig. Jennifer scowled. "I'm going to sausage that sow. I hate that pig." This was the kind of thing Jennifer always said about White Pig, but because she reliably produced large litters, she was kept on, her volatile temperament notwithstanding. (Until the spring of 2004, when she killed most of a litter of seventeen piglets, and then she was shipped.)

The boar they had rented from Mark Durkee to impregnate the sows had never done what he was supposed to do—had just lain around listlessly, eating and drinking, showing no interest whatsoever in procreation. Kyle and Jennifer finally called Mark Durkee to come and get him, which he did. He killed him in the pen and took him away to turn him into sausage.

"We fed him forever," Kyle said.

"Four months," Jennifer said. "We got screwed on that one. We're getting another boar from Patty and George Gast, the people who also owned Buddy the Bull. We're getting a six hundred-and-fifty-dollar boar for free. In return, we give them two piglets a year for three years. The disadvantage is five months of feeding it when he's not doing anything."

The talk turned to the markets. "As far as going to Norwich, I want to go," Jennifer said.

"I didn't know that was on the table. That's where we sell things," Kyle said with some exasperation. "Lebanon, no."

Brad, who was looking through the sidelights by the front door, exclaimed suddenly, "The turkey's eating the pumpkin!" Not a turkey, but a goose, which was wandering around by the wagon that held the pumpkins.

Jennifer and Kyle forged ahead. They decided not to participate in either the Lebanon market or the smaller South Royalton market, except as an occasional, unreserved vendor. Rather, they would concentrate their efforts on selling to the co-ops, particularly the South Royalton market, where they had earned $3,600 this year. Should they increase the potatoes, which they had sold in their entirety? They had sold out of onions, leeks, and shallots in September, and if they planted more next year, they could have some for the 2004 Christmas market at Norwich's Tracy Hall.

The garlic was so labor intensive, Jennifer said, that they should keep it at the same level.

They argued over the summer squash. Who picked more, Jennifer or Kyle? Kyle insisted he had picked a lot of pattypan squash. Jennifer wanted to plant more summer squash: they were easy, they needed almost no care, they sold out, and they could be direct-seeded into the ground, rather than being started in the greenhouse. The most labor-intensive, she said, were the cucumbers and the beans, which, because they grew so large so quickly, had to be picked every few days, a process that took hours of their time.

"And the flowers," Kyle said.

"We already agreed we weren't going to do flowers, except for CSAs," Jennifer responded, nettled. "What if we do cucumbers for CSAs and us? We got $107.06 from SoRo and at least that much from Norwich. For CSAs we had cucumbers every week from July seventh on."

"It seems like a lot of work keeping up with them," Kyle said.

"How about growing them on platforms?" she suggested. He resisted.

"Well," Jennifer said, taking a deep breath. "How do *you* want to grow them differently?"

"Direct seed."

"*I* think we should do direct seed on the summer squash," Jennifer said, returning stubbornly to her earlier point. "The pumpkins and melons were grown in the greenhouse. Maybe we didn't do winter squash in the greenhouse. All the pattypans we grew in the greenhouse, and they took off. I think we should have less intensive vegetables in cultivated land, and interplant the corn and the pumpkins."

"Interplanting doesn't work all that well," Kyle said.

"I disagree," Jennifer said.

"I don't."

"I know you don't because it's *your* statement."

They didn't know whether to look at each other or away, to leave or to stay. Both were on the point of tears, or murder, or both.

"Look at all the gourds. They did great," Jennifer said. Kyle gave her a look that was somewhere between resentment and frustration that his points did not seem to be taking hold.

"Yeah, we've got gourds," he said, pointing out the window to the wagon, where the pumpkins and gourds sat, "and they're sitting down there rotting. "

"Where do you want to plow onions?" Jennifer asked.

"I don't want to plow more," Kyle said.

"Debba has two bathrooms!" Brad interjected, out of nowhere.

"We did a good job interplanting with buckwheat. . . . So what do you want to do?" Jennifer said doggedly. This devolved into a discussion of which crops could be covered with green, or raw, manure—buckwheat, oats, or peas—and where they should go, which devolved into further tension.

"I'm getting a lot of resistance, so we won't do it," Jennifer said. After some further talk, she accused him of being too negative. "I don't want it to go like this," she said.

"I'm sorry I was being negative," said Kyle, the words dragged out of him, clipped and expressionless.

"I'm saying if we take the same area and use less-intensive crops, then it will be less labor intensive," Jennifer said, "but you're saying, you don't want to do it."

"We don't *need* to grow melons, we don't *need* to grow flowers."

"I know. I already took them out," Jennifer said, nearly yelling. "So we're growing less corn?" (Kyle loved corn, loved to grow it, loved to tend to it, and to eat it; a nostalgia perhaps for his Midwestern childhood.)

"However—it—works—out."

"We can also use the space for pumpkins which we don't believe in either," Jennifer said shakily.

Bouquets of cut flowers for sale at the Lebanon Farmers' Market. The farm had the space to grow them in the greenhouse, but they didn't sell well, and won't return.

"I have to go to the bathroom," Kyle said abruptly. He pushed his chair aside, and stalked off and was gone for five minutes. When he returned, the climate between them had altered, and they made a concerted effort to talk more evenly and neutrally.

"What about CSAs?" Jennifer asked.

"I've always been opposed to increasing them," Kyle said. This led Jennifer to brood over the people who she felt were querulous, never satisfied. One family issued complaints weekly, it seemed. The woman who confronted Jennifer at market, saying her lamb was simply too fatty. Too fatty! And what was Jennifer going to do to make restitution! The woman who seemed irritated that she wasn't getting lamb when she thought she had paid for it. All the work-share people who, predictably, began the summer with the best intentions in the world, and then faded away as the work kept coming and coming.

"Well, all I know is, I don't need grief," Kyle exclaimed, and whether he was talking about shareholders or Jennifer wasn't clear. Brad sat in his rocking chair, its rails making cricket-chirping noises as it moved back and forth on the floor. Jennifer did another tally with the calculator.

"The CSAs," she declared, in an Aha, I knew it voice, "got $759 worth of produce for paying $600."

"The CSAs have a lot of turnover anyway," Kyle observed. "It's a strange way of getting food."

"Last year we made a clear profit from it. This year it's not that trackable. Three work-shares petered out. What if we did it on a different day?" (By this she meant not on a Thursday, when they were also getting ready for Saturday's Norwich Farmers' Market.)

"Well, we've eliminated farmers' markets," said Kyle, "with the exception of Norwich."

Jennifer noted that she was asking one of the work-share people to be responsible for collecting and tracking the money in 2004. This year, although prepayment had been required in March, most of the shares had not been paid until closer to summer, and some hadn't been paid until well into summer, and there were a few who still owed them money. In future, they would send bills of remittance out to late payers. No more shrugging it off. They couldn't afford to shrug it off. To make the shares more cost-effective, Jennifer posited, they could reduce the number of $14 chickens in the full shares from ten to eight, and offer turkeys at a reduced rate.

"Why are we offering them bargains?" Kyle objected. "I thought you were complaining about them getting too good a deal."

"The greenhouse is the best deal we have," Jennifer observed.

"I just don't want to waste a lot of time growing stuff that doesn't work out," Kyle said.

"We had extenuating circumstances with the melons; the woodchucks ate them. Speaking of eggplants," Jennifer said, in a non sequitur, "How much do you want to grow?"

"As much as you want; just don't make me eat it," Kyle groused, but now he was joking. Both of them detested eggplant, just as both of them detested runny eggs; of such minutiae is a partnership made.

"I want to start the greenhouse earlier than February sixteenth," Jennifer said.

"You're welcome to start seeds two weeks earlier. *I will be in the woods!*" Kyle stated.

"The only other thing is the apprentices, notwithstanding haying. Are we going to have two or three apprentices?"

"Two. Because you always add one anyway."

"Well, then, we need to make it more educational for them, because we won't be paying them as much."

"That's what *I* said, and *you* said . . ."

Jennifer, sharply: "You know what I mean. We need to make it more attractive for them since we won't be paying them as much."

"I don't think we need to dock people's pay if they go to workshops."

"I did not do that," Jennifer protested.

"Yes, you did."

"I did not."

A trio of cows—from right, Buttercup, Andrea, and Blackberry Stem—enjoy the waning days of fall. At the first heavy snow, the bovines are put into the barn for the winter.

Kyle then reminded her that two apprentices in 2002 were docked $3 for going to a workshop.

Jennifer sputtered with indignation. "Yesssss," she said, and reiterated for Kyle the offenses that had led to the pay docking. The apprentices in question had, one evening, kept Jennifer awake until three in the morning by banging around ostentatiously in the kitchen, preparing food to take with them to a workshop on organic gardening. To compound this offense, for which they did not apologize, they had then installed all the jars of food in the refrigerator with signs that read, DON'T TOUCH. As if Jennifer or Kyle would take their food! As if these apprentices weren't being given a free place to stay and free meals!

This out of the way, Jennifer continued, "If we have two apprentices, one will stay in the house. The general idea is, do you still want to have apprentices?"

"Yes, I wouldn't be agreeing to all these vegetables if we weren't going to have apprentices," Kyle said equably.

"I could ask Russ to help on pipeline," Jennifer said. She sighed when talking about Russ. "He's twenty-six and he doesn't know what he wants to do."

"*Shame on him,*" Kyle said sarcastically. "I'm forty-three . . ." He left the implication unsaid.

"Kyle!" Jennifer erupted.

"Jennifer!" Kyle snapped back.

They retreated.

"So you won't be in the greenhouse with me soaking up the warmth?" Jennifer asked plaintively.

"Of course I will. I just want you to know where my priorities are, and they're in the woods."

They looked at each other, and in the look was exhaustion, resignation, familiarity, a grudging respect and affection. Kyle joked that they would have been divorced by now if it weren't for the fact that she'd lacked the ambition to file the papers. She laughed; it was a relief to be able to laugh, to dissipate the tension.

"Cows?" Kyle said brightly.

"I didn't bring cows up," Jennifer said, frowning.

"I know. I am." Kyle looked at her, eyebrow raised in challenge.

"I am *keeping my cows!*"

The cows remained, but by early December, Jennifer and Kyle had decided to send Bobby, their last draft horse, down to Massachusetts, to Ben's family's farm. Ben had expressed an interest in continuing to work with him over the winter. The idea was that Bobby would return the following summer, although they both knew they were only indulging sentiment when they said this. Really, this was a precursor to selling him. Bobby stood idle, when he should have been plowing or carting. Sometimes he seemed bored, racing around the pasture because he had nothing else to do. He had picked up fairly quickly on what Ben was trying to teach him this summer, and would benefit from the discipline.

Ben and Meg came to get him on December 11, driving up in a brand-new, $12,000 trailer, the cost of which Ben was sharing with his two brothers. (Jennifer and Kyle would do some exclaiming between themselves over the newness of the trailer, and the cost, and the apparent ease with which Ben and his brothers had purchased it, which was a far cry from their cadging equipment wherever they could get it.)

December had been quite mild so far, in the 30s and 40s, and it was warm enough this day so that it was raining. The hills behind the house were veiled in fog. Water dripped off the eaves of the house and the barn. There was snow on the ground, maybe a foot, and it was dissolving to slush in the rain. The turkeys were gone; they had been slaughtered the Friday before Thanksgiving in Weybridge, near Middlebury, and both Jennifer and Kyle were relieved to see the last of them.

Thanksgiving had been an amiable affair. Jennifer's parents, and Anna and Rick, and Gene and Laura Craft, had wrapped around two long wooden tables that stretched from one end of the house almost to the other. Jennifer had spent weeks poring over the twenty-three-page instructions on how to make a "turducken"—a chicken stuffed inside a duck inside a turkey. Jennifer had been as obsessed with that turducken as Kyle was obsessed with his wooden bowls. Anna had helped her bone all the birds, a process that took six hours, and helped her load the turducken—an enormous, misshapen affair, stuffed to overflowing with bread crumbs and chestnuts and prosciutto—into the oven, where it cooked for nearly twenty-four hours at 150 degrees.

On the table had been homemade bread and homemade butter, cornbread stuffing, brussels sprouts and butternut squash and mashed sweet potatoes, all from the garden. There had been American champagne and wine and water. They'd served, in lieu of traditional pies, cheesecake with a berry sauce. There were candles, and plates of different patterns, because they didn't have enough of the same plates to go around. Jennifer and Anna had both worn long dresses, with princess sleeves and snug, shirred bodices and flowing skirts. Jennifer's was purple, and Anna's red. It was the only dress, Jennifer said, that she owned, and it was years old. Jennifer had her hair down and brushed to a high sheen, as did Anna.

Kyle gave thanks for Brad and for the meal. Jennifer gave thanks for Brad and for her family and for the harvest, which had put this food on this table. Laura gave thanks for Gene and for fourteen years of marriage passed more or less peacefully, without murdering each other. Louis Megyesi gave thanks for his wife and his children and grandchild. Gene gave thanks for Laura and for his friends. Anna gave thanks for Jennifer. And then they ate and drank until it was all gone, and Kyle came over to stand behind Jennifer and put his big hands to her face and brushed back her hair and allowed as how he guessed she had outdone herself this year.

Now Thanksgiving was behind them, and Christmas was not far away, and then, just after the New Year, Brad's fourth birthday. Last year, Kyle and Brad had driven out to Circleville for Christmas while Jennifer stayed home to watch the animals. No such trips this year. They'd already been to Circleville in October, and Jennifer's parents had come for Thanksgiving, and they figured that their familial obligations had been discharged until next year.

Ben and Meg had not been back to the farm since they'd left three months earlier, and they hugged Jennifer and Kyle. They trooped into the horse barn, where Michael and Bobby were still in their stalls. Jennifer would use Michael as a kind of lure, to draw Bobby out to the trailer. Bobby had left the farm but once, when he went down to Brattleboro, to be green-broken, or trained. He was not used to trailers and he was not used to leaving the farm. The sum of his world had been some nine acres of pasture and one narrow horse stall.

Because it was raw and damp, they were going to put a blanket on him for the two-and-a-half-hour ride to Massachusetts. Bobby didn't like blankets as a rule, chafed at the way they sat on him awkwardly, and disliked the straps that went under the belly. They didn't have a blanket big enough for him, so he was going to wear Michael's. Jennifer went into the stall and began to maneuver the blanket around him. "Look," Kyle said to Brad, whom he was holding in his arms. "Bobby's getting a blanket on him and doesn't even care—yet."

"You're a handsome fella," Ben said soothingly to Bobby. "Those cows aren't going to know what to think."

Bobby stood still, but in his eyes there was nervousness. Why were all these people standing around him, and why was Jennifer putting a blanket on him, when she knew he hated them, and why was everyone staring at him? He shifted his head from side to side, as though to get a better picture of what was happening.

"Mom's trying to buckle that blanket underneath him, but it's too small. It's Michael's," Kyle said to Brad.

"Good boy, Bobby, that'll make you nice and toasty," Jennifer assured him, once the blanket was secured.

"You look handsome," Ben told him.

"Do you need any brushes?" Jennifer asked Ben.

"If you've got some."

From the empty stall opposite, she retrieved three brushes, a pick for cleaning feet, a coarse comb to get the mud out, and a currying comb.

"Are you going to be a good boy? Going to go play?" She kissed Bobby's nose. She turned to Michael, and haltered him: "Okay, Michael, you're the decoy."

Brad gets a kiss from Bob, a Suffolk Punch draft horse. Kyle and Jennifer intended to use the rare breed horses to haul lumber, collect maple sap, and harrow the garden. With parenthood, however, they weren't able to do as much with them, so Bob was loaned out to a former apprentice to continue working.

Jennifer led Michael, and Ben, Bobby. They backed them out of the stalls and led them out of the Horse Barn, and down the muddy slope and up to the gate that separated the pastures from the gardens. They led them through the gate and up a narrow dirt path to the road and around to the trailer. Jennifer easily brought Michael up into the trailer, and hitched him.

Then followed a prolonged and intricate courtship, in which Jennifer and Ben tried to cajole the deeply reluctant Bobby, his head bowed stiffly in resistance, to walk up into the trailer. Jennifer talked softly to him, and so did Ben. Look, they told him, Michael's already in the trailer. See? It's okay to go in, too. They promised him that everything would be all right, and they meant it in the larger sense, but Bobby was no fool and he knew they were trying to get him to do something that he didn't ordinarily do, the implications of which, from his point of view, were not good.

They tried the grain bucket, then a water bucket. Bobby sniffed but retreated, his powerful haunches doing a delicate backpedal. Whatever this soft soap was they were selling, he wasn't buying. Kyle stood on the snowy slope that led to the driveway, hands in pockets, watching. Cody, the dog, trotted around busily, but then lay down, sinking into the snow; he, too, watched. Brad had a sled that he was pulling up and down the slope; he paid intermittent attention to what was going on by the horse trailer.

Jennifer led Michael away. His presence was not much use, and was proving distracting. Meg made a big show of putting hay in the trailer, and held a clump out to him. Again, Bobby lifted his head to smell. He looked interested, but he stayed firmly rooted in place. Now Jennifer tried hitting his flanks gently with a rope. Meg stood behind him with a whip and touched it to him. Bobby would not move.

"Can you work at the back?" Jennifer asked Kyle.

"I'm working with Brad right now," Kyle said. Jennifer gave him a look.

As if sensing that the attention had shifted away from him, Bobby shied up and made a break for it, but was quickly hauled back in. Meg handed Ben the whip. But really it was now between woman and horse, Jennifer and Bobby.

"Good boy, good boy," she intoned. "Come on, come on. Up a step, up a step." She crooned to him. *Good boy, good boy* became a chant, made powerful by its soothing repetition.

Bobby did a little two-step, a sort of jiggly tap dance. "I know, I know," she said sympathetically, "but you're not going to get your way, buddy."

Bobby kept raising his legs and putting them down, raising them and putting them down. He knew what was expected of him, but he was afraid to do it, like a child who has been dared to do something that involves risk to himself, but also some shame if he doesn't follow through.

Jennifer bribed him with grain again. "Come up, you can do it." A UPS truck rattled by. Bobby stuck his nose into the trailer.

"He'll probably put his foot up there, and scare the crap out of himself," Jennifer observed. Prescient words, as Bobby chose that moment to relieve himself. He shifted with some agitation, then put his front legs hesitantly onto the step. "Good boy, good boy," Jennifer murmured. Then the back legs, and he clattered into the trailer.

"Close the door, close the door, close the door," Jennifer yelled to Kyle. Bobby was backing up, thinking better of it. Kyle's first instinct, as always, was to locate Brad safely—who was oblivious to the drama and was occupied shoveling snow—and that done, he quickly latched the door. A minute later, Jennifer and then Ben emerged.

Bobby was now inside alone, and he whinnied and made the trailer heave and shake with the stamping of his feet—an earthquake that continued for some minutes, until Ben and Meg climbed into the cab of the truck, started the engine, and carefully worked the trailer into such a position that it could turn safely onto Morse Road. With the trailer swaying, partially because of the uneven road and because of its size and because Bobby was rocking it back and forth, it crawled down the road.

Jennifer watched the trailer lumber away, with something like regret on her face. She loved Bobby, loved his good-natured high spirits, but that wasn't a good enough reason to keep him, even if he was one of the first animals born on the farm, with obvious sentimental value. Sending Bobby away marked, in more than one sense, the end of a chapter in the story of this place. Relinquishing him was just one of the numerous accommodations Jennifer and Kyle had had to make to keep the farm running. It was not that they had abandoned idealism for hard-nosed pragmatism, exactly, because their farm was still fueled largely by ideas and ideals, but that their ambition had inevitably been tempered by reality.

"We both had more income before we started farming. We had $42,000 in the bank, in joint savings. Kyle had $180,000 in pension retirement, a 401(k)," Jennifer said later, sitting at the table, as she went over the past year with him. "We were going to buy a house, not a farm. We had a very clear focus on buying a house. We put $11,000 down on the house right away. We spent $16,000 the first year on livestock."

"I hope you're wrong," Kyle moaned.

"Do you know how much the horses cost us?"

"I don't want to know."

"We had no concept of what poverty was," Jennifer said, shaking her head, marveling at their naivete. "We picked up these rare breed horses."

"I wanted to have draft horses and use them," Kyle said. He pointed out the window. "We had one foal and he left this morning."

"Kyle quit his job and Brad had all those medical expenses. It was July and I realized we were not going to have any money at all. Now people call us up and ask for advice. Which blows my mind."

"People who know us don't," Kyle said with a short laugh.

"Any type of self-sustenance you do is farming," Jennifer argued. "Why does it have to be all or nothing?"

"As long as we're married and we're here, we'll be doing it," Kyle said. "If we're growing food for ourselves, that's still farming."

"Self-sustenance is a form of farming," Jennifer reiterated. "But we want to better figure out how to fine-tune."

"I want to consolidate," Kyle said. "If we get big somewhere, get smaller somewhere else." How would he consolidate? "Oh, like, get rid of beef cattle, for example, or the horses."

"We need to get a pony for Brad," Jennifer said quickly.

"That's not consolidating," Kyle pointed out.

It was, Kyle concedes, "a painful summer, emotionally: work, the relationship. A frenetic, day-to-day pace, never being caught up." The only reason the farm made it at all, he says, was because they could borrow and barter.

"It's not that we couldn't buy a tractor, it's that we refuse," Jennifer said.

"NO DEBT," Kyle boomed.

"It incenses me to have debt," Jennifer pronounced.

"We don't plan to take out a loan," Kyle said.

Even so, the bartering was, Jennifer admits, delicate. "It takes a lot of bargaining."

"This year we didn't get our second cut of hay, which was a pretty big deal," Kyle said.

"But I think people are starting to get it," Jennifer noted. "George and Agnes started by demanding cash. Last year we raised a pig for them. This year, they got half a beef. But you have to network carefully, and sometimes you can get bitter."

The core of their existence, and perhaps the larger reason why they farmed—even with all the financial uncertainty, the lean periods, both financially and emotionally—was Brad. For all the challenges of running the farm, and the marriage, they both were certain that the farm was a better place to raise their child than almost anywhere else. From it he would learn, they believed, the values of patience, perseverance, and hard work; respect for the land and the animals, and respect for family, friends, and neighbors. That was the long view. Of course, whether Brad would appreciate it when he got older and was expected to share in the chores is an open question. "He hates

Brad helps Kyle plant garlic cloves. Planted around Columbus Day, the roots of the plant grow in the fall and the cloves are harvested in late summer.

some of it," Jennifer said. "He hates sheep. He told me, 'When I grow up, I want to live on a paved road.' But I think it's good for him to have that experience and stay out of trouble."

Kyle was skeptical. "I don't think farming necessarily correlates with staying out of trouble."

Although neither might freely admit it publicly, because it would involve a softening and a vulnerability where they were often used to skirmishing—a skirmishing that seemed to serve the purpose of moving them forward, of pushing them toward something, even with the one sometimes biting angrily at the other's hand—they were together for reasons that went beyond the fact of their son.

"Other men didn't want to do farming," Jennifer said, by whom she means the men with whom she had been previously involved. "They didn't want a moral life. I can't explain it. We had such a community in Maine, and it just felt like the only way we were able to get *that* close to people was to grow food. I wanted that kind of closeness with people."

All those years, apart and together, in which they had studied and observed and lived with the birds, the animals, the vegetation, the water, the sky, had made them understand the world in a certain way. It was not everyone's way, but it was theirs, a shared perspective without which they would be more likely to drift apart. For themselves, for the farm, for their son, they wanted the same things. They had the same instincts on how to live a life that was worth living, as they saw it, and had arranged their lives accordingly.

Kyle, who was by nature of the glass-half-empty school, liked to say of the farm that there was "no money at the end of this tunnel." But there was something larger, that harkened back to what Crèvecoeur called "a great exuberancy," and that was the gratification of taking the farm, which had lain dormant, and watching it slowly, slowly, reemerge.

They had carved out from the land an existence that with each passing year seemed less tenuous, more rooted, and more orderly: yields had increased, livestock had increased, cultivated land had gone in area from what Kyle called a "postage stamp" to two acres. The resources of the farm were managed as a whole, not as discrete entities which had only a vague relation to each other.

The property looked lived-in again; animals and people roamed where, for a while, there had been nothing. It had been a farm in 1872 and in 1900, and it had been a farm during the Depression and World War II and the Korean War and the Vietnam War, and only for a short while, really, had it *not* been a farm. But in that brief period, there had been a marked decline, which was no one's fault, it was just the way it was. Time had done its work, just as time and circumstance had contrived to turn all those hill farms into ghosts, cellar holes and stone walls cut into landscape where hardly anyone went anymore.

It had not been easy to halt and reverse the farm's decline, but Jennifer and Kyle and Brad, and everyone who had ever helped them, had done so. You could call it a labor of love, but that would sentimentalize it. Labor, yes; love, not always. Tenacity, conviction, determination, begging, bartering, guile, charity, and pragmatism and idealism in equal measure: these were the things that had brought them through, and secured their future, and in a sense, their liberty and their autonomy.

There were bonds between people, but there was another bond that went as deep, even if it was often unconscious; the bond between farmer and land, which was not unlike that between parent and child. It often tried one's patience and taxed one's energies, but it was a deep companionship and understanding and knowledge that arose out of simply being together, nothing else, and from the realization that the bond was intentional and deliberate, not arrived at lightly, and one meant to last for years.

Five years earlier, Jennifer had asked the Morses what they would think if she and Kyle painted the farmhouse yellow, instead of the white it had always been, and although they didn't say no outright—because, technically, they were no longer the owners—she could tell they were taken aback and disapproved somehow, that in some ways they still thought of it as theirs, and were reluctant to see it change. This year, when Jennifer had asked them about painting the house, they'd had no strong reaction. "They didn't care," she says.

Fat Rooster Farm now belonged, in every conceivable way, to Jennifer and Kyle and Brad—and they, to it. Little wonder they cherished its possession.

With their gardening tools in hand, Jennifer and Brad walk back to the house after planting onions.

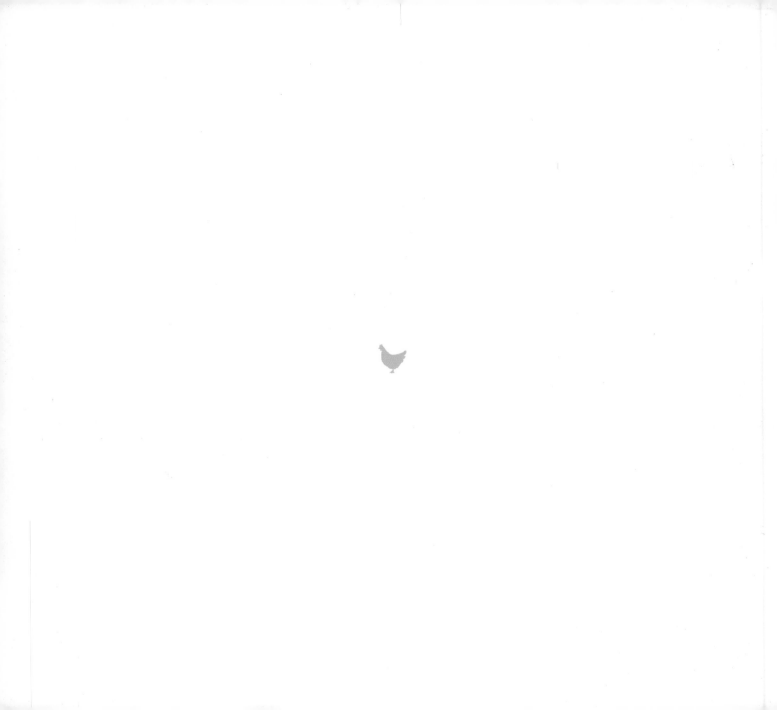